PERMANENT MAGNET TORQUE HARVESTING

JAY A LUNKE

Copyright © 2020 Jay Lunke

rights reserved.

ISBN:
ISBN-

DEDICATION

This book is dedicated to Jesus Christ who has had a large influence in my life on how to conduct one's life. I am thankful for the forgiveness of sins and the promises He has given me about the new life I now have in Him. My wife and I wrote the book "Please pass the PEW" in love for Jesus Christ.

CONTENTS

Introduction:

Permanent Magnet Torque Harvesting

What is Permanent Magnet Torque Harvesting?

Why create a functional magnet through reconfiguration?

How does reconfiguration occur without physically moving parts?

How large of a permanent to electromagnet ratio does it take to achieve over unity status?

Five to one, torque from permanent magnets to electromagnets

Summary of motor Application

Power Generation

The big question is why use an indirect method to generate electrical power?

Building a prototype motor for the hobbyist

Hobbyist Corner

Failed All permanent magnet build

About the author

Conclusion

Disclaimer

ACKNOWLEDGMENTS

I have had great teachers in high school and in vocational school who did a great job in explaining the technical aspects in the area of electronics. But the greatest education and learning are the life lessons that can be only learned from prayer for understanding. I have always prayed for understanding at home and at work in how to do things the correct way. God does not get enough credit from people who get answers from Him as they pray. I know that if I did not pray, that I would not have the ideas that I have written in this book.

INTRODUCTION:

My name is Jay Lunke. I have been fascinated with magnets since I was a child and I still am as a senior citizen. As I would play with magnets, I had always wondered how to use the pull they have to each other in a motor design. I, like many people, have tried different "all permanent motor" designs that ran into the sticky point problems. After several design failures, I started working on designs that had more permanent magnets than electromagnets with the thought that you have to pay for the electrical energy to operate the electromagnets but not the permanent magnets. I came up with the flow through motor that basically has the electromagnets flowing through the middle of the permanent magnets. Everything needed to be custom made and after getting the quote for the permanent magnet build, I put the design on hold until I came into money to build them. That never happened. So, I now work with designs that people can build with off the shelf magnets. I have developed a new technology that uses up to five permanent magnets to one electromagnet with all positive forward torque through the full rotation of the rotors, having no sticky points. Also, the motors have amazing power so that they will be able to replace the magnetic motors in vehicles and get three times the range. But that is not all, I also have designed different power and control circuits. I power the electromagnet circuits with the electromagnets being one leg of a tank circuit. I use steering diodes to control the direction of the current of the circuit so that the capacitor in the other leg of the tank circuit does not need to operate at the resonant point in the circuit in order to provide high efficiencies in the motor. The modified circuit collects the back EMF into the capacitor in order to use that energy the next time the electromagnet is used again in the motor. It is the combination of these two technologies that I believe I am harvesting enough torque from the system in order to power a generator that will have power left over after powering the motor. I will go step by step through these technologies in this book. I show you many applications. I provide plans for a prototype motor you can build with a parts list of off the shelf parts. Building this prototype motor will provide a test bed that will allow you to improve on motor designs using this new technology. Building a prototype motor is the best way to learn how this new technology works and how to enhance it for unlimited applications.

What is Permanent Magnet Torque Harvesting?

It is a method of getting permanent magnets to use there forward torque to produce work by operating in the Three Layer Electromechanical Movement Technology without having reverse torque in the movement (no sticky point). We all can feel the torque from a magnet when we are holding a magnet in our hand and bring the magnet close to a steel object. The magnet will pull itself to that metal object. In order to pull the magnet off the metal object, it takes as much force in the opposite direction to accomplish that task. So, it has been thought by most people that it is not possible to harvest more torque from the magnet than the initial movement of the magnet as it pulls itself to the metal object. After you read how the Three Layer Electromechanical Movement works,

I hope that you have a different understand of how permanent magnets can produce work through the usage of the torque they have within them.

What is Three Layer Electromechanical Movement Technology?

Let's say you have a gravity wheel that is bolted to the ground with one weight on the wheel. To start with gravity operates normal as we know it today. As the weight moves to the right side of the wheel, the weight pulls the wheel to a clockwise direction. when the weight moves to the bottom of the wheel; the normal gravity will prevent the weight from moving up the other side. Now let's say you found a way to change the direction of gravity so that weighted objects fall to the sky instead of the ground. You then make that change and now the wheel continues to move from the bottom to the top by the weight on the wheel. Then you change the gravity back to normal operation again. By switching this control of gravity direction will create a continues operating motor.

What this new technology does is to move the north and south poles of the magnets in the stator assembly that then provides a constant positive torque between the rotor and the stator assembly causing positive torque in the motor through the full rotation of the motor.

This technology creates mechanical movement between two objects having magnetic components on them by reconfiguring the magnetic components on one or both of those objects. These two objects are usually a rotor and a stator assembly. This reconfiguration can be physical movement of physical components, but is more likely to be done by creating a functional magnet being made up of permanent magnets with an electrical device like a coil or electromagnet. This technology will not function without creating a functional magnet through reconfiguration. A functional magnet in this technology is a magnet built with the electromagnet with the two adjacent magnets to it, that together perform as one larger magnet. The magnet in the middle of the two adjacent magnets, I call the switching magnet. The switching magnet can be either a permanent magnet or electromagnet. When using electromagnets in the reconfiguration, it is a functional reconfiguration, not a physical reconfiguration. In fact, the technology cycles between two states operating in four segments of movement between the two objects. There will be a lot of demonstrations coming up in how this happens and how the torque harvesting occurs.

The reason I named this new technology into Three Layer Electrical Movement Technology is because I see two-layer technology as a rotor operating with a stator assembly. I consider the third layer of this technology as the stator assembly or rotor continually changing its configuration back and forth during the operation of the motor in order to provide the torque harvesting as it operates.

WHY CREATE A FUNCTIONAL MAGNET THROUGH RECONFIGURATION?

The normal characteristic of two magnetic objects passing each other is that about 50% of the interaction with each other is positive torque and 50% of the action is negative torque. So, what the Three Layer Electromechanical Movement does is to use the 50% positive torque between these two objects "as is" and then reconfigure one of the two objects in order to change the normally negative torque into desired positive torque direction. It does take energy to perform the reconfiguration process using this technology. After the reconfiguration has ended, the device goes back to the normal configuration at a position where the permanent magnets are able to use their built-in torque to create more movement. So forward torque is produced at all times with the technology even though external energy is only required for 50% of the movement. When you compare this to conventional motors, they require power on the motor all of the time.

Now this 50% power reduction is the minimal savings in capturing torque from a permanent magnet in order to perform work because I have designed motor devices that have torque movement ratios of 5 to 1. This is torque from five permanent magnets and one electromagnet used to generate movement in the device. This is like having a long boat with six sets of oars in it and you only have to pay for one man doing the rowing because the other five people are rowing for nothing. This paper will get into details of how that later can happen on. Now re-configuration of the switching magnet can occur either by physically repositioning hardware or without physically moving parts when turning power on to an air core electromagnet that is in the switching position of the device.

How does reconfiguration occur without physically moving parts?

Now a Copper wire without electric current flowing through it is not attracted to a permanent magnet. So, when you wind the wire into a coil having an air core, also the coil is not connected to anything, the Permanent magnet has little to no effect from it. Now when you put a current through the coil of wire, it creates a magnetic field that interacts with the permanent magnets in close proximity to it. The Three Layer Electromagnet Movement Technology takes advantage of these characteristics in the design of the devices. The following drawing shows this new technology as it uses the reconfiguration in motor designs.

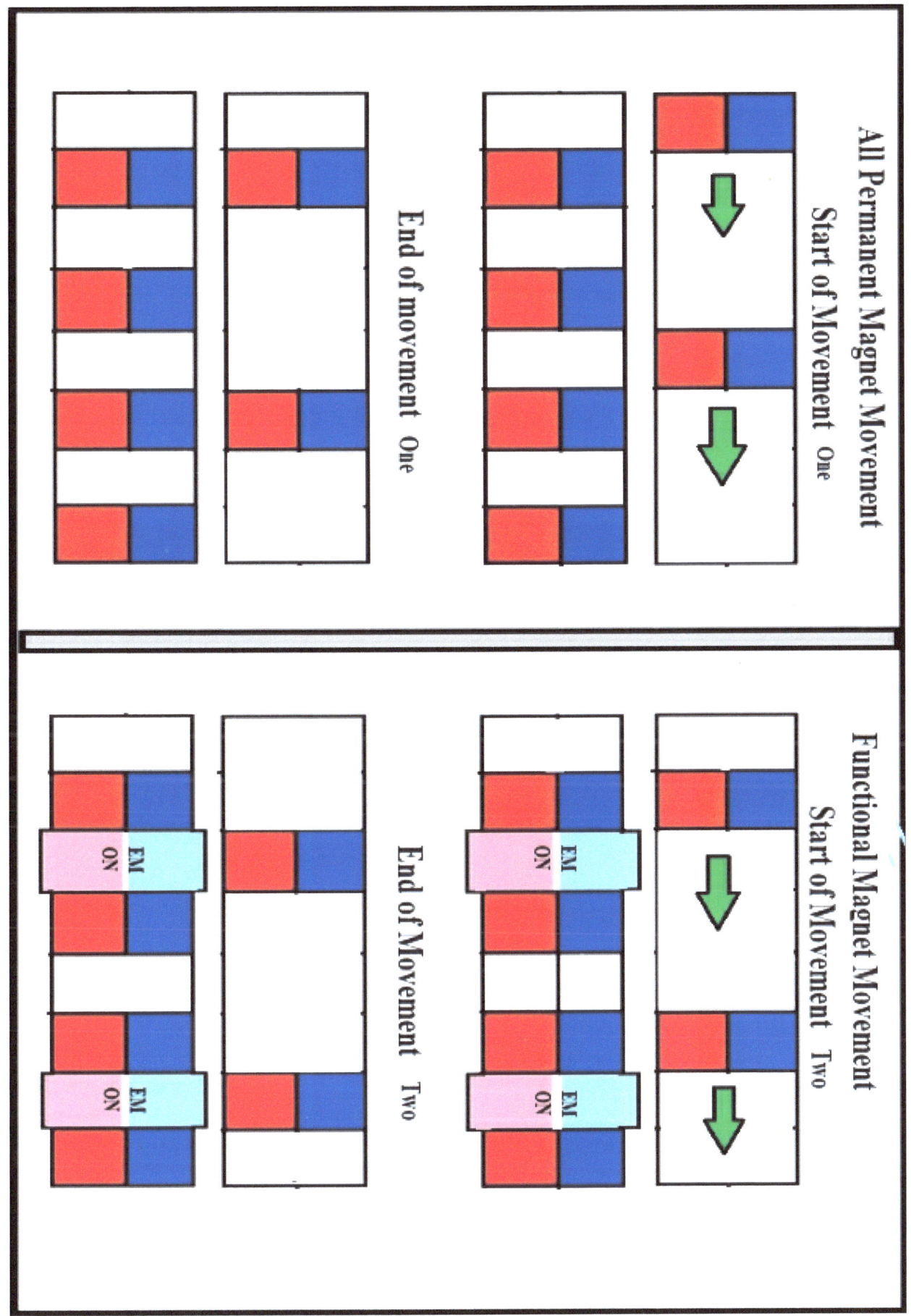

The all-permanent magnet movement in movement one, has torque without any outside force. It is the torque that is built into each of the permanent magnets that pulls the magnets to align with each other. So, this free movement becomes very important in the Three Layer Electromechanical Movement Technology. Now I did not draw the electromagnets in the all-permanent magnet movement drawing, because the power is off and they do not have a metal core and they are not connected to a circuit like a generator coil would be. The torque during this time will be produced from the interactions of the permanent magnets in the two assemblies. Again, these two objects would normally be a stator and rotor assembly.

Now the right side of the drawing has the reconfiguration that is done by powering every other electromagnet. The electromagnet along with the two adjacent permanent magnets next to it creates the functional magnet. This functional magnet changes the center points of the poles in the functional magnet to the center of the powered electromagnets. Now the blank spot in the stator assembly has an electromagnet with the power off. So, this technology uses two sets of electromagnets each having a 25% duty cycle. Each circuit is on its own switching cycle. Notice that I have the electromagnets larger than the permanent magnets. I have done this by design in order to ensure that the electromagnet has more power with the other assembly in order to ensure the movement of the magnets in that assembly in order to move that assembly into alignment with the electromagnet. Now when you look at the functional magnet, then you will see that only every other electromagnet is turned on. This creates better efficiency because each set of electromagnets will only have a 25% duty cycle. Let's look now at the next two movements of the two objects.

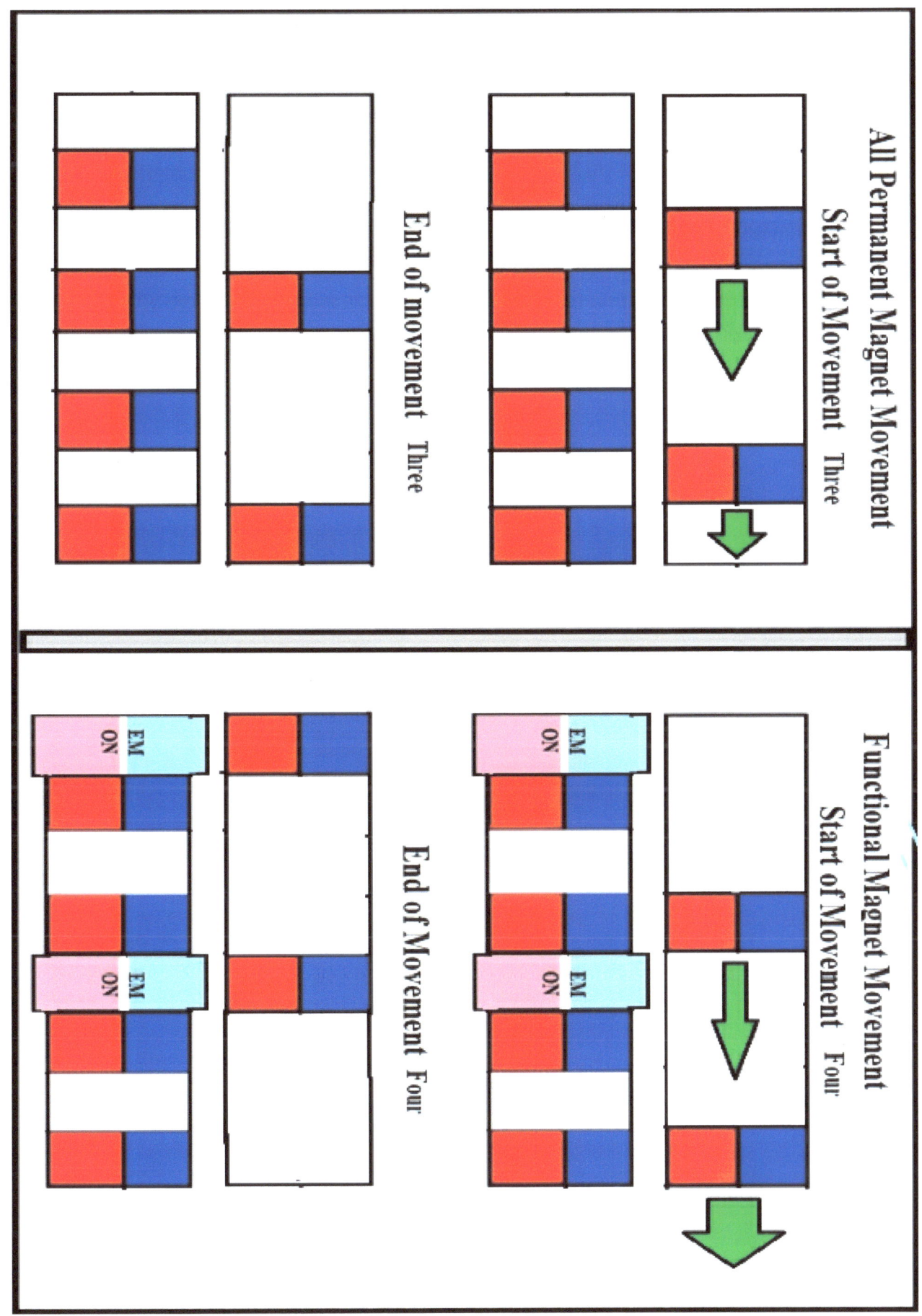

These are the third and fourth movements used in this new technology. The third movement uses (all permanent magnets) for the torque movement again. This action is the same as the first movement. The fourth movement works like the second movement. It uses a different set of electromagnets for the movement. The electromagnets that were off during the second movement are now on and the ones that were on are now off. So, this circuit of every other electromagnet are operated at a 25% duty cycle. This reduction of electrical energy demand in the electromagnets means better efficiency. These magnets will operate cooler because they will not heat up as much as conventional electromagnets in today's magnetic motors.

Let's look at the following drawing in order to compare the torque in electric and magnetic motors to the torque that can be used from permanent magnets using this new technology given in this book.

Three Layer Electromechanical Movement technology

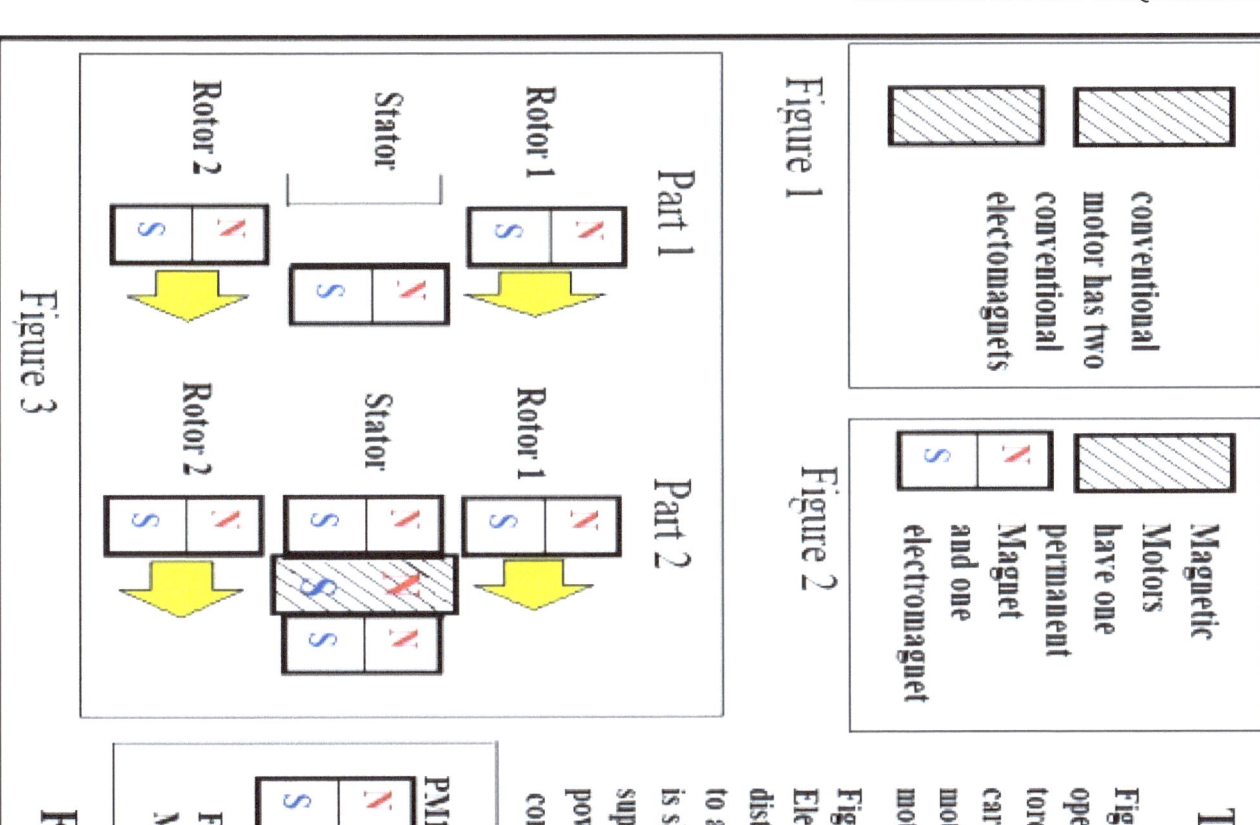

Figure 1

Figure 2

Figure 3

Figure 4

Figure 1 and 2 show what I call two layer electromechanical movement. It is a rotor operating with a stator. Figure one shows the electric motors that are built with torque with all electromagnets. Figure 2 shows magnetic motors currently used in cars having one permanent magnet per one electromagnet. You have magnetic motors in electric cars over electric motors because you get better range from those motors.

Figure 3 shows the two of the four segments of travel in a Three Layer Electromechanical Movement technology. Both part 1 and part 2 move the same distance in a motor. Part 1 does not require any electricity in order for the magnets to align each other through attraction. In part 2, the electromagnet is activated as is shown in figure 4 in order to produce a functional magnet. When the power is supplied to the electromagnet the functional magnet is activated and when the power is turned of it gone as far as the magnetic fields go because the coil has an air core.

Part 1 and Part 2 repeat themself with a different electromagnet circuit in parts 3 and 4 in order to complete the four segments of travel of the rotors moving through the stator assembly. When you add up the torques from each parts of Part 1 and Part 2, they add up to torque from 5 PMs to 1 EM. Even if you caluculate less than I do, the torque ratio is better than the current magnetic mores used in cars today. This technology needs to be explored. Please help me out in bringing this technology into reality.

Jay Lunke 12-15-2020

There have been valid concerns from people about the usable torque in using this new technology.

In movement Part 1, at the start of the movement, the rotor magnets are balanced between two stator magnets. So, at the start of the movement there is not forward movement because of this balance. It is as the rotors move away from the backside magnet and starts to come into more contact with the forward magnet, that the forward magnet has more pull than the magnet on the backside. So, in my opinion if you did a vector analysis through the full start to stop of the movement that the forward magnet will have a total of three times the torque than the backside magnet on this movement. Now remember that there is no external power needed for this movement because it is created by the torque in the permanent magnets. All of the electromagnets are powered off during this time.

Now in Part 2 of the movement, a functional magnet is created consisting of an electromagnet with the two adjacent permanent magnets. Now the electromagnet is closer to the rotor magnets in order to have dominance over the permanent magnets in the functional magnet.

Now there is a give and take here that needs to be better understood through testing. That is that the larger the distance the stator permanent magnets are from the rotor magnets, the electromagnet will not need as much current in order to build enough movement to align the rotor and stator functional magnets together. But the farther away the stator permanent magnets are away from rotor magnets, then the smaller the torque will be in movement part 1. So, testing needs to be done to optimize the distance of the permanent magnets between the stator and rotor assembly in relationship to the electromagnet to rotor permanent magnets. Later in this book, there is a chapter in building a prototype motor that has adjustable permanent magnets in the stator assembly in order to optimize the motor for this.

In my opinion, even with this design option of not having the full torque that can be taken advantage of with this new technology, having a larger percentage of torque harvested from having a ratio of five permanent magnets to one electromagnet will still produce a motor that requires less electrical energy per horse power than a magnet motor. I want you to know that without performing the testing, the design theory in my opinion will perform two to three times better than the magnetic motors in vehicles today. There are other magnet configurations we will be looking at later on is this book, one of which will likely perform better than this current configuration.

These four movements in the new technology can be used for linear or circular motion. One of the best examples of a 3 to 1 permanent to electromagnet torque motor design is in the following torque motor example. Now this motor example is very basic and there are much more productive designs coming later in this paper. There are other disk motor designs, but None that use this new technology that I know of.

With smaller devices, the first two movements can be used in the designs. The two stage devices use the permanent magnet for the common mode operation of the device. The power going to the electromagnet is used to produce the none common movements. An example of this is the operation of a magnet that moves scrapped cars. The permanent magnet would lift and move the car or truck with no power to the permanent magnet. The electromagnet would only come on in order to drop the car at its new location. This saves a lot of power in this operation. There are so many more applications of using these first two stages of operation. There are unlimited designs and applications that save power in these devices.

The following drawing is a simple design using this new technology.

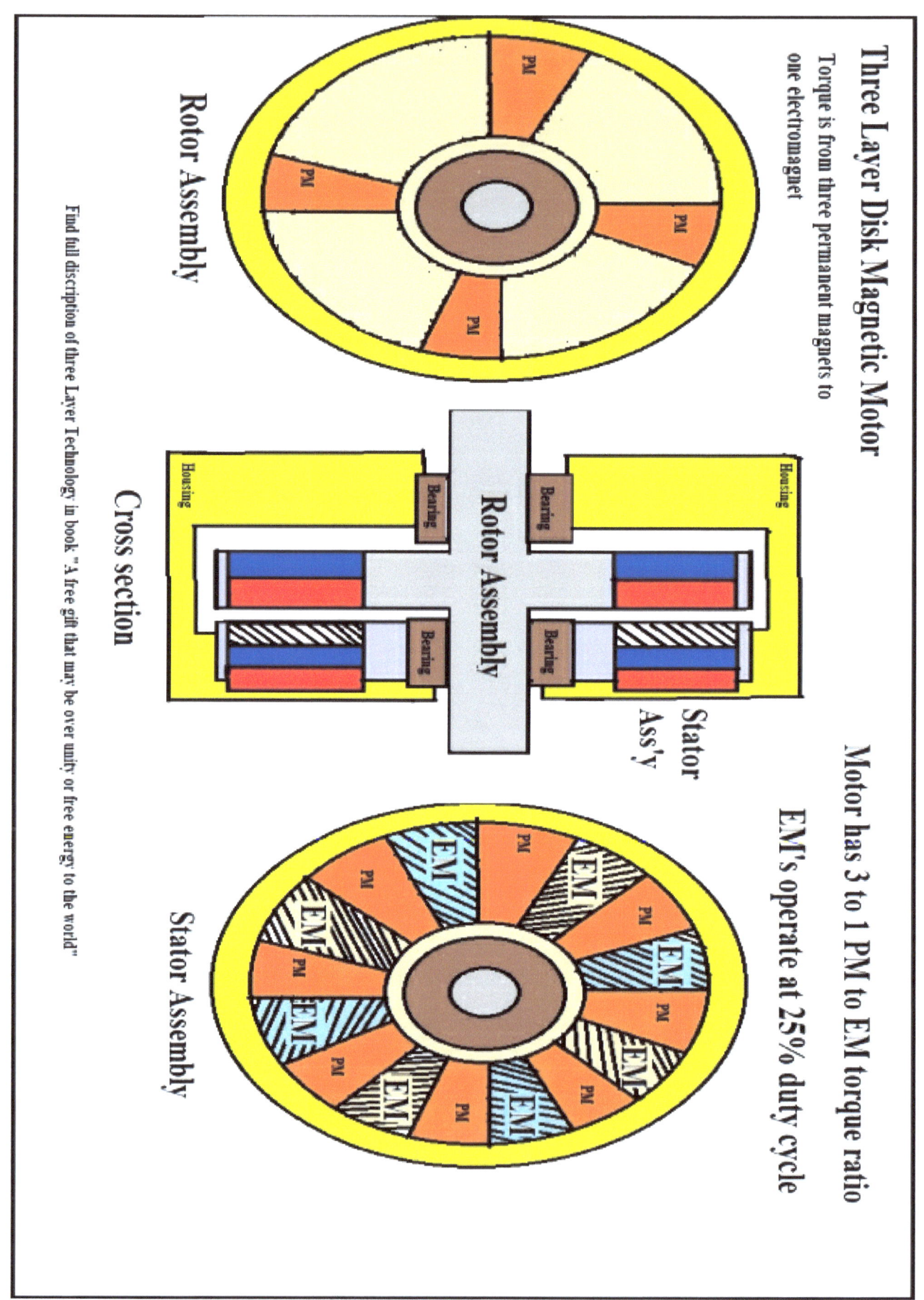

Now the two electromagnet circuits are shaded with different colors. With the disk magnet configuration there is the ability to stacking them into different layers which places more power into a smaller package. The biggest limiting factor with this new technology is that the rotor permanent magnets can only be placed in one of the four segments of travel in order for the technology to work properly. If you place more permanent magnets in the rotor, this will reduce the efficiency of the motor to the point that it may not function at all. So more is better does not hold true with this technology. This means other motors can produce more power for the size of the motor. Later in this book I show designs using 5 permanent magnets per one electromagnet. So this 3:1 design is more for displaying the technology. This technology has efficiencies that will provide much further range in cars and trucks and other transportation.

Now permanent magnets produce torque in a motor and electromagnets produce torque in the motor. With this technology, the torque from the permanent magnets are free and you only have to pay for the energy for the electromagnets used in the functional magnets that produce torque in the motor. It makes sence that with more permanent magnets producing torque than electromagnets, that there must be a point that there will be enough torque from the motor in order to power a generator that would feed power to the motors electromagnets and still have power left over to power other things. While magnetic motors have a ratio of 1 to 1, this motor having a ratio of 3:1 is much better. Now when you see the 5:1 motor designs, you will wonder why they are not in cars and trucks today.

How large of a permanent to electromagnet ratio does it take to achieve over unity status?

The reason that electric vehicles use magnetic motors is that fact that having magnets in the motor design are superior to all electromagnet motors. So, it only makes sense to increase that ratio in order to bring larger ranges to those vehicles. The drawings we have looked at so far, have a torque ratio of three to one, torque from permanent magnets to electromagnets. So, what can be done to design a motor with a greater torque ratio than that? The next section will show you that.

I see that a magnet motor that uses a ratio of one to one, permanent to electromagnets is the most efficient motor currently used in vehicles today because of the fact that they are actually harvesting torque from the permanent magnets making them that good. Now when performing a vector analysis of the permanent magnet to electromagnet actions will most likely show that the individual magnets have less than a 50% conversion to do work in the motors. I believe that is why our current magnetic motors are not self-running with a COP>1. Now just think of increasing the permanent magnet to electromagnet ratio will have on the work that is performed by the motor for the electrical energy to operate the electromagnet. But the goal does not have to be self-running, If the range is only doubled, then this new motor design makes sense over the current magnetic motor designs.

Five to one, torque from permanent magnets to electromagnets

Permanent magnets have torque within them that do not require any external power to operate them. Remember this new technology only operate the permanent magnets in forward torque modes. The electromagnets are the only components in the motor that use power outside of the control circuit. The control circuits power consumption is low compared to the electromagnets power consumption. The 3 to 1 torque ratio motor designs use two electromagnet circuits each operating at a 25% duty cycle. The 5 to 1 torque ratio motor design use the same electromagnet configuration using a 25% duty cycle in order to power the electromagnets. The way this can happen is that the stator has been slightly modified with wider electromagnets that extend beyond the width of the permanent magnets in order to be able to interact with two rotor assemblies instead of one rotor used in the 3:1 torque ratio design. As you can see in the drawing that since the number of electromagnets stay the same, the addition of another rotor that only has permanent magnets on it will increase the permanent magnet to electromagnet ratio.

Note:

Since electromagnets have a north pole and a south pole, we need to take advantage of both poles in our designs. This is doubling the work performed by the electromagnet. The way I do this is by having two rotors working with one stator. The rotors are all permanent magnets, while my stator is built with 50% permanent magnets and 25% electromagnet circuit one and 25% electromagnet circuit two.

There are a lot of designs out there that could improve their performance just by using both sides of the electromagnet into the design. Look for them and you will find them.

> Now it is time to take a close look at the 5:1 torque ratio functionally. The reason I say functionally is because we want to only look at the invisible flux interactions between the stator and rotor assemblies because that is what creates the torque between the two assemblies causing the movement in the motor.

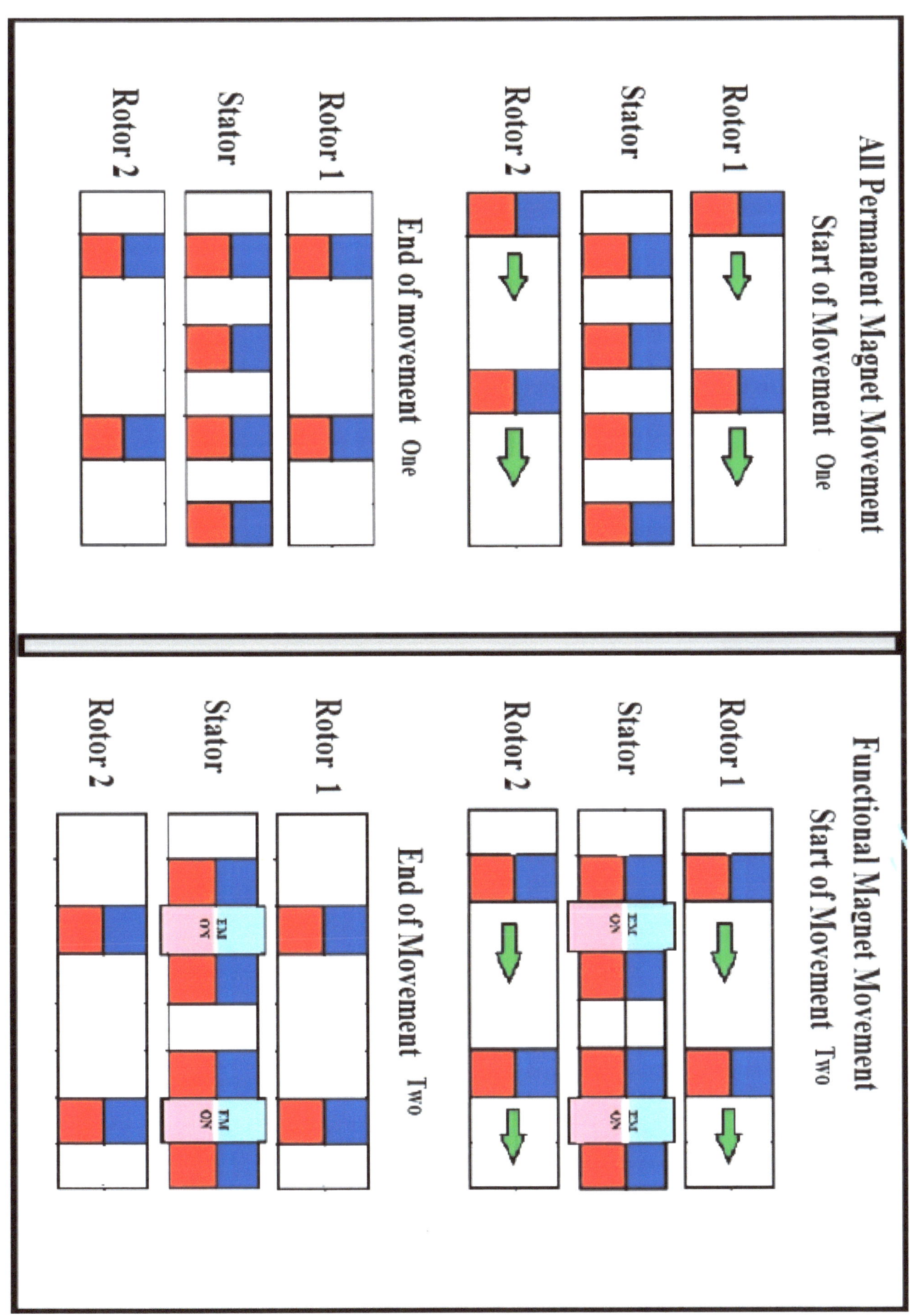

The torque from permanent magnets is like turning a marry-go-round in the playground. There are some trees close to the marry-go-round. So a child on the merry-go-round can push on the tree to increase the speed of the merry-go-round. A child on the ground can push on the railings on the merry-go-round in order to increase the speed of it. Now a magnet on the rotor can pull on the stator magnets in order to provide torque on the motor. At the same time the magnet on the stator can pull on the magnet on the rotor to produce torque. Now on my motor designs, I use attraction rather than repulsion for the rotation. I feel this is easier on the magnets in doing this.

Now when you look at the drawings above, movement one has the rotor one magnet pulling on the stator magnet. You have the rotor two pulling on the stator magnet. Then you have the stator magnet pulling on the two rotor magnets. So I see this as a three to zero torque movement. Zero being the fact that there is no power to the electromagnets. This is free torque movement in the motor. Now in the second movement, you have rotor one permanent magnet pulling on the stator functional magnet. You have rotor two permanent magnet pulling on the stator functional magnet. Then you have the stator functional magnet pulling on the rotor permanent magnets. This is a two to one PM to functional magnet ratio. Since the functional magnet has an electromagnet in it that requires about the same power as the permanent magnet has, this is about the same as a 2:1 ratio for the part 2 motor movement. When you add up movements one and two together, you end up with torque from five permanent magnets to one electromagnet. Segments three and four have the same ratio using a different electromagnet circuit. In fact since the four movements repeat themselves over and over again, the final motor assembly has a 5:1 ratio.

Now in the magnetic motor, the torque does not have a direct force of torque to motor movement. A vector analysis shows that the torque to motor movement is less than 50% (this is a rough estimate, my opinion). This is why the magnetic motor has not reach over unity status because the electromagnet and permanent magnet torque to movement force is less than 100%. Now with my 5:1 torque ratio, if each magnet contributes 25% of its torque into motor movement, then the five permanent magnets and one electromagnet would have a 150% torque to motor movement conversion. This is only my opinion and theory of how this would work. Only building and testing with third party verification can this be proved out.

Since this is a new technology, I want to show the four segments of the technology again in a different step by step format. This format is much more informative.

Permanent Magnet Torque Harvesting

Movement One of Four Movements

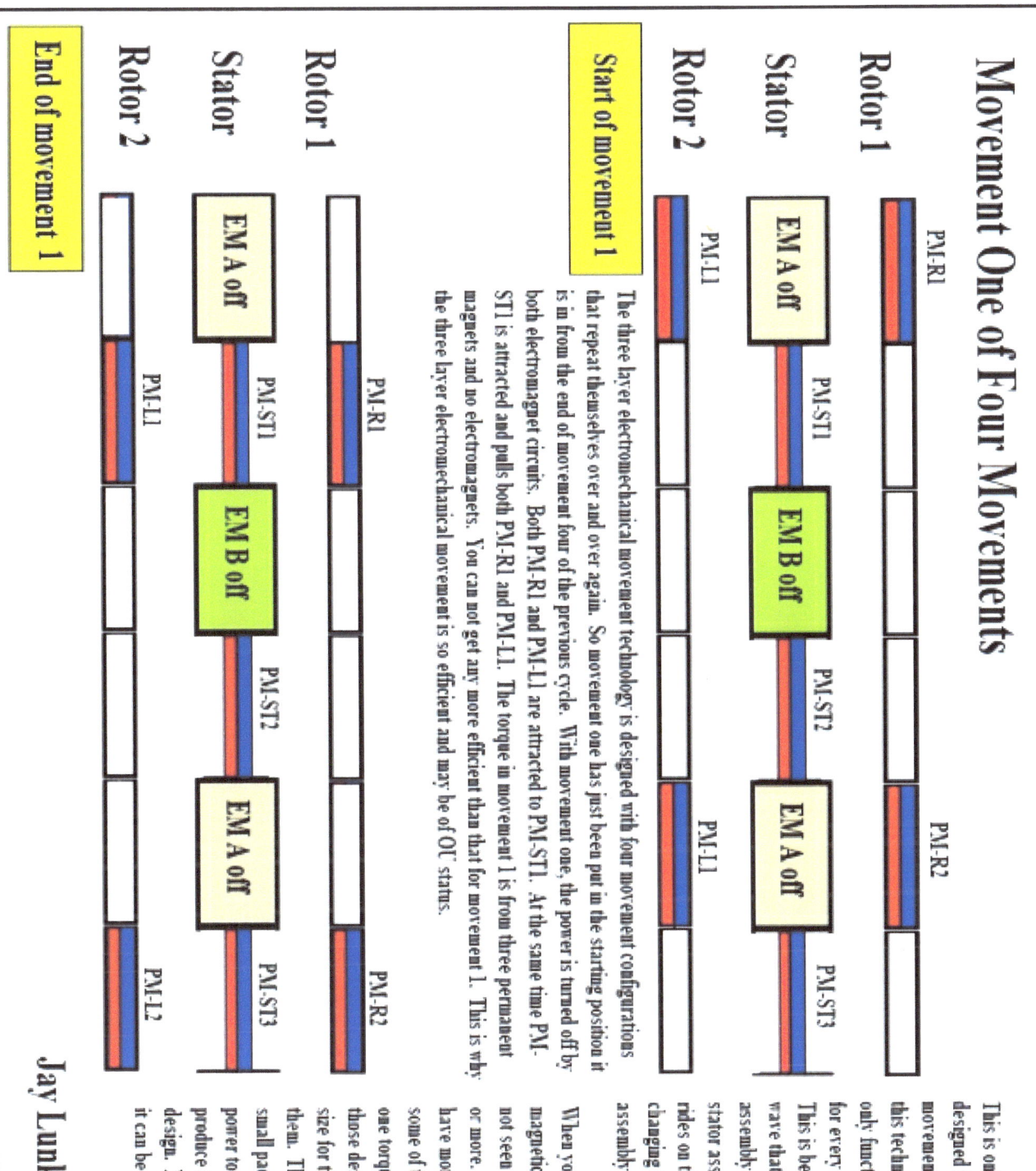

The three layer electromechanical movement technology is designed with four movement configurations that repeat themselves over and over again. So movement one has just been put in the starting position it is in from the end of movement four of the previous cycle. With movement one, the power is turned off by both electromagnet circuits. Both PM-R1 and PM-L1 are attracted to PM-ST1. At the same time PM-ST1 is attracted and pulls both PM-R1 and PM-L1. The torque in movement 1 is from three permanent magnets and no electromagnets. You can not get any more efficient than that for movement 1. This is why the three layer electromechanical movement is so efficient and may be of OU status.

This is only the first of four movements that is designed into the three layer electromechanical movement. All four movements are needed for this technology to function. This technology can only function with one rotor permanent magnet for every four possible positions on the rotor. This is because this technology creates a flux wave that continually travels around the stator assembly; by continual reconfiguration of the stator assembly. The rotor permanent magnet rides on the crest of this wave through the changing torques between the rotor and stator assembly permanent magnets.

When you compare the torque from electric and magnetic motors on the market today, I have not seen any that have a torque of five to one or more. There have been some designs that have more magnets to electromagnets, but some of them have multiple magnets to create one torque from the collection of them. In those designs, the motors can become large in size for the amount of torque you get from them. This design offers a lot of torque for a small package with minimal required external power to produce that torque. This should produce many applications for this motor design. Now there are restrictions of how small it can be made.

Jay Lunke Modified 12-7-20

Movement two of four for New Technology

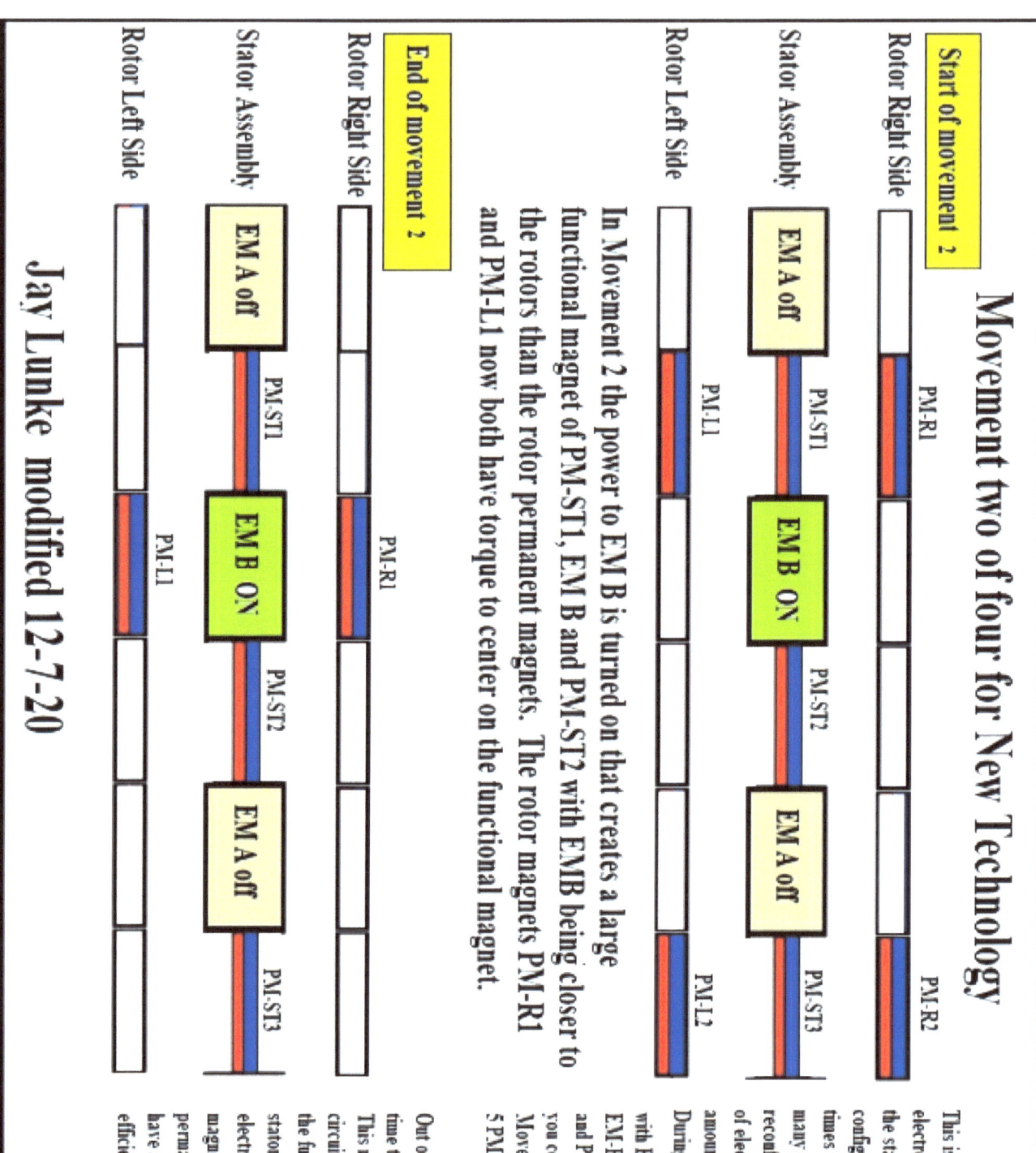

In Movement 2 the power to EM B is turned on that creates a large functional magnet of PM-ST1, EM B and PM-ST2 with EMB being closer to the rotors than the rotor permanent magnets. The rotor magnets PM-R1 and PM-L1 now both have torque to center on the functional magnet.

This is a unique function of the three layer electromechanical movement is to reconfigure the stator function by changing the functional configuration of the stator magnets several times during the operation of the motor using as many permanent magnets during the reconfiguration in order to reduce the number of electrical magnets needed in the total amount of torque produced in the motor.

During movement 2, PM-R1 has torque to align with EM-B, PM-R2 has torque to align with EM-B. EM-B has torque to pull both PM-R1 and PM-L1 into alignment with itself. When you combine this with the three torque of Movement 1, you end up with a torque ratio of 5 PMs to 1 EM magnet.

Out of the four movements this is the only time that power is applied to the EM-B circuit. This means a duty cycle of 25%. The EM-A circuit is off during movement 2 which allows the functional magnet to be created in the stator assembly. Now most conventional electric motors do not have permanent magnets. Magnetic motors have a 1 to 1 permanent magnet to electromagnet ratio, so have a ratio of 5 to 1 puts this motor way more efficient than other motors.

Jay Lunke modified 12-7-20

Movement Three of Four for new technology

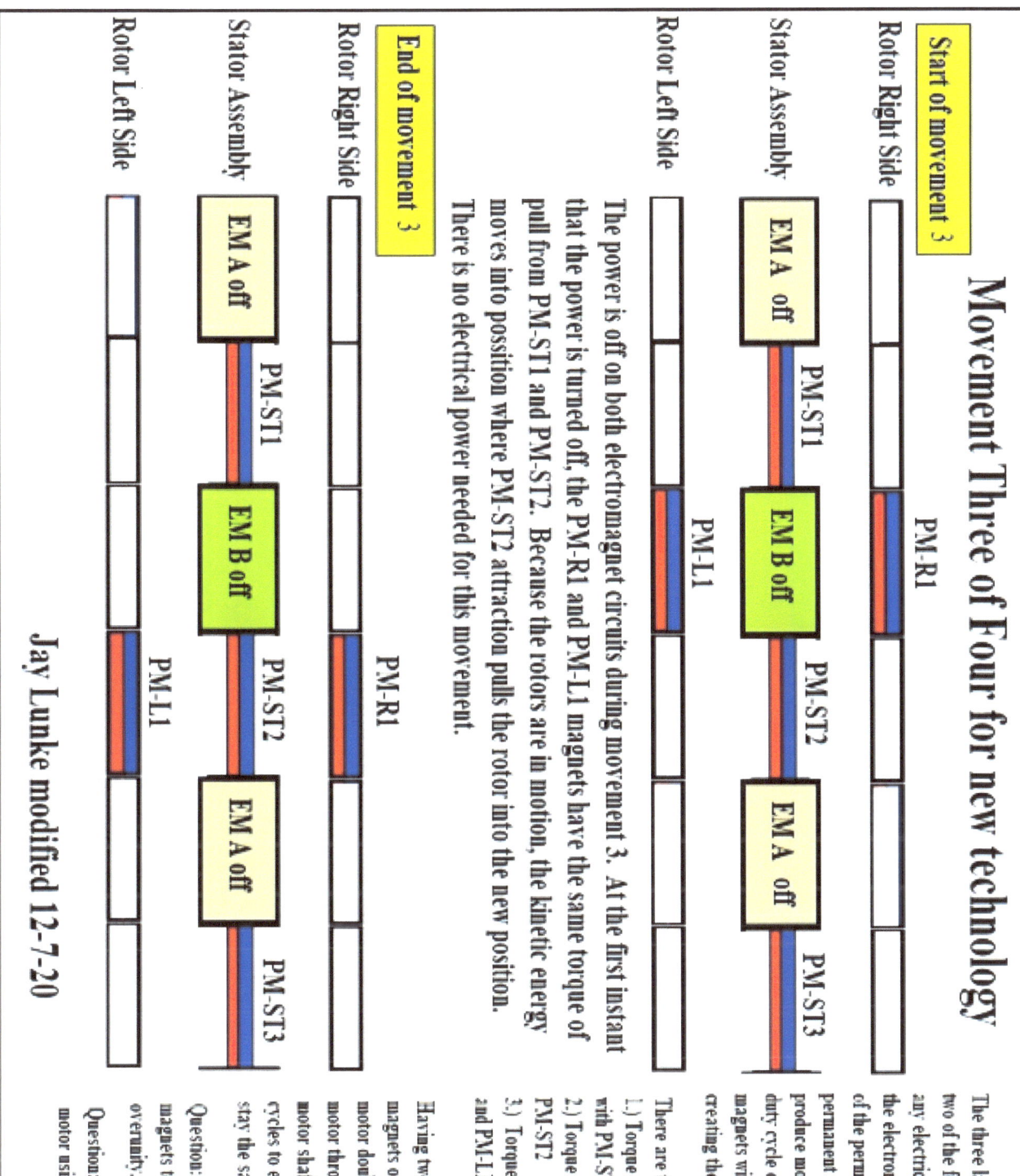

The three layer electromechanical movement has two of the four movements that operate without any electrical power because when the power of the electromagnets is turned off, then the torque of the permanent magnets interacting with the permanent magnets in the rotor assemblies produce movement of 22.5 degrees to occur. The duty cycle of of having only the power off to the magnets with torque from the permanent magnets creating the rotor movement is 50%.

There are three torques in movement 3.
1.) Torque from PM-R1 pulling itself to align with PM-ST2.
2.) Torque from PM-L1 pulling itself to align with PM-ST2
3.) Torque from PM-ST2 pulling on both PM-R1 and PM-L1 to align with itself.

Having two rotors with only permanent magnets on it providing forward torque in the motor doubles the power of the motor of the motor through the doubling of torque on the motor shaft. With the second rotor, the power cycles to each of the electromagnet circuits stay the same.

Question: At what torque ratio of permanent magnets to electromagnets will produce overunity.

Question: How much more efficient is this motor using a tank circuit with steering diodes.

Jay Lunke modified 12-7-20

Movement Four of Four for new technology

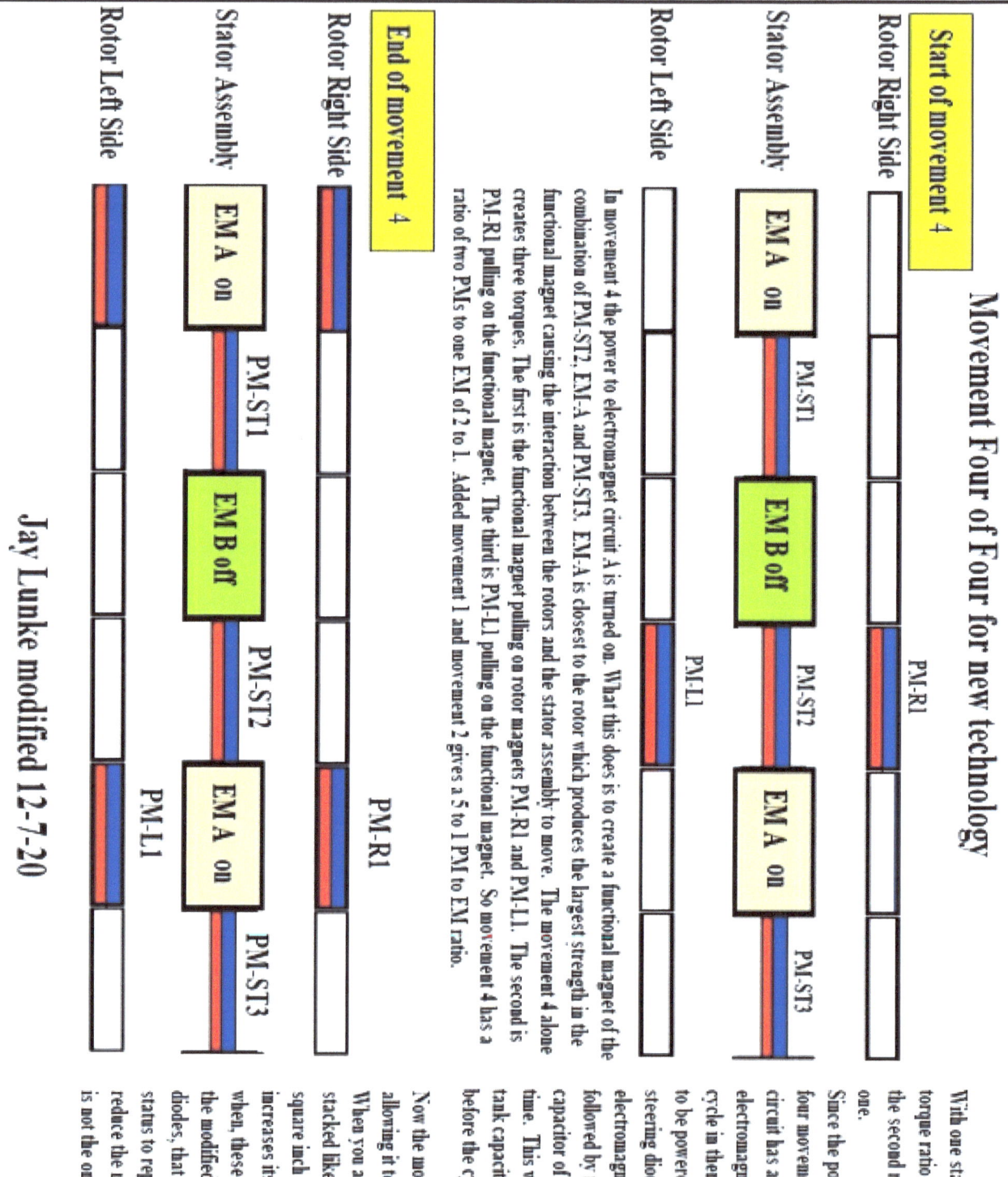

In movement 4 the power to electromagnet circuit A is turned on. What this does is to create a functional magnet of the combination of PM-ST2, EM-A and PM-ST3. EM-A is closest to the rotor which produces the largest strength in the functional magnet causing the interaction between the rotors and the stator assembly to move. The movement 4 alone creates three torques. The first is the functional magnet pulling on rotor magnets PM-R1 and PM-L1. The second is PM-R1 pulling on the functional magnet. The third is PM-L1 pulling on the functional magnet. So movement 4 has a ratio of two PMs to one EM of 2 to 1. Added movement 1 and movement 2 gives a 5 to 1 PM to EM ratio.

With one stator and one rotor, the PM to EM torque ratio would be three to one. By adding the second rotor changes that ratio to five to one.

Since the power to EM-A only occures in one of four movements means that the electromagnet circuit has a duty cycle of 25%. Because the electromagnet circuits have only a 25% duty cycle in them means that they are good circuits to be powered by the modified tank circuit with steering diodes. The circuit would power the electromagnet circuit for 25% of the time followed by recovering the back EMF into the capacitor of the tank circuit for 25% of the time. This would be followed by topping of the tank capacitor for 50% of the remaining time before the cycle repeats itself again.

Now the motor configuration is compact allowing it to be used in many applications. When you add the fact that this design can be stacked like pancakes adds a lot of power per square inch of the motor assembly. This increases its applications a lot. If, I believe when, these motors are built and operated with the modified tank circuits using steering diodes, that this system will be meeting OU states to replace many current systems to help reduce the need for fossil fuels. Of course this is not the only OU option in the works.

Jay Lunke modified 12-7-20

The reason that this technology is torque harvesting is because if you replaced all the permanent magnets with electromagnets, you would have a motor that has the same performance at the cost of a lot more energy in order to operate it. So, the difference in energy to operate this motor is what is harvested from the permanent magnets. This torque is changed into work that is performed in the motor. Now if you get more work done with less electrical energy used to perform that work, torque has been harvested in that process. In my opinion, there has to be a point in the magnetic motor design where the ratio between permanent magnets to electromagnets becomes great enough to run a motor that feeds a generator and have power left over. So, the magnet acts like a battery, instead of providing electrical energy like a battery, it produces torque that can be converted into work.

Now that you have a better understanding of how the technology works, it is time to look at applications of this technology in different designs.

The following 5 to 1 torque ratio of PM to EMs shows the motor design in a simple drawing. Adding hardware to these drawings will not help, but hinder the purpose of showing the operational theory of the Three Layer Electromechanical Movement Technology in applications.

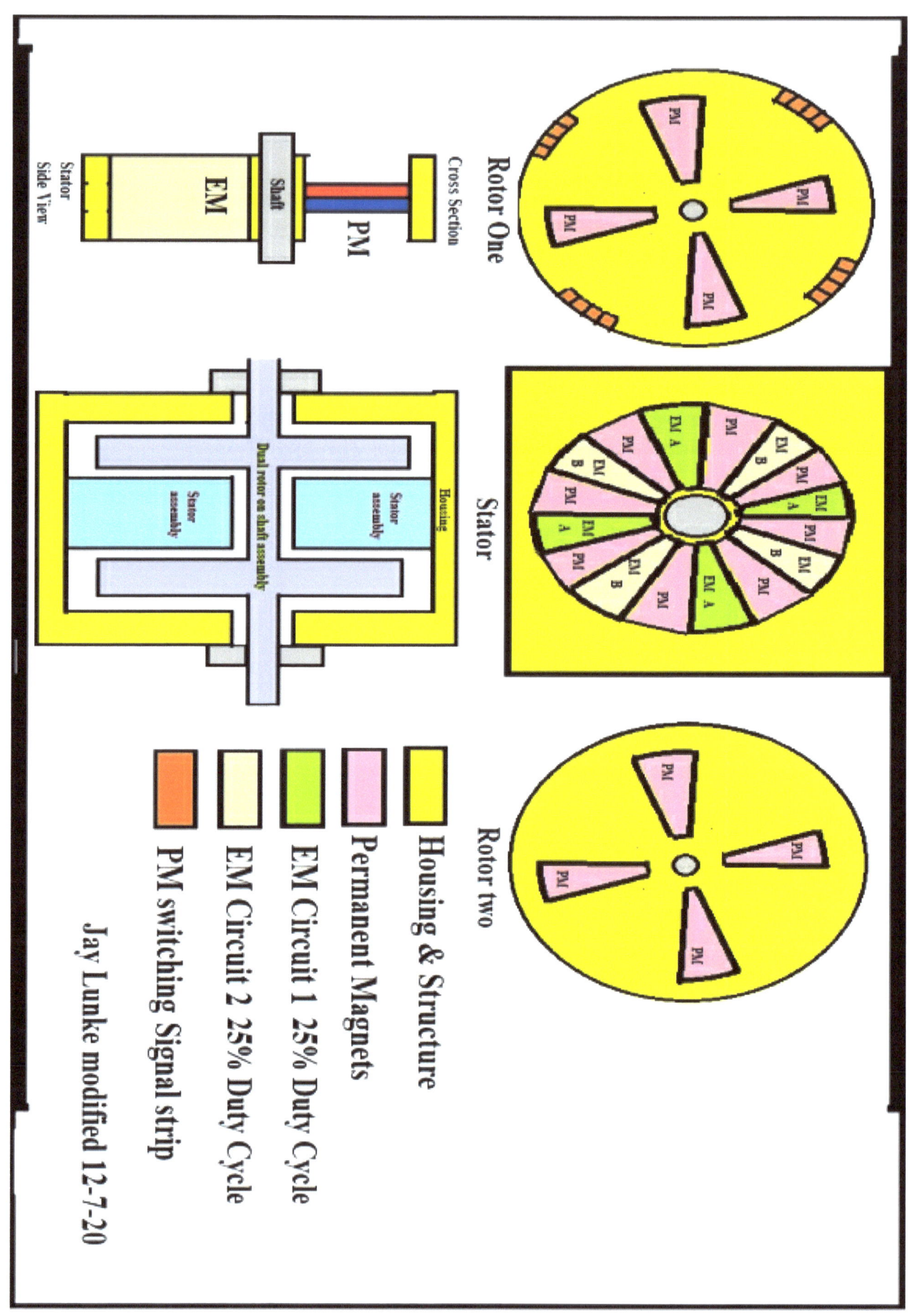

Jay Lunke modified 12-7-20

Let's compare this motor design to current magnetic motor designs used in cars today.

Today's auto magnetic motors use a 1 to 1 ratio of PM to EM torque ratio. The electromagnets operate at a 100% duty cycle rate. My 5 to 1 PM to EM torque motor uses a 25% duty cycle on each of the two electromagnet circuits which will operate without creating as much heat. The efficiency of my motor will require a lot less electricity per horse power generated from it. Now the current magnetic motor may have about twice the power per cubic foot of motor. Now that means more room will be needed under the hood for my motor design. The current magnet motors are already smaller than gas engines, so my motor will not be larger than those gas engine and transmission dimensions.

The following motor design stacks the disks. The following motor design is equal to five motors. The package is only two to three times larger than a single motor. In my opinion, this design could produce torque results that would be similar to the magnetic motors placed in cars today.

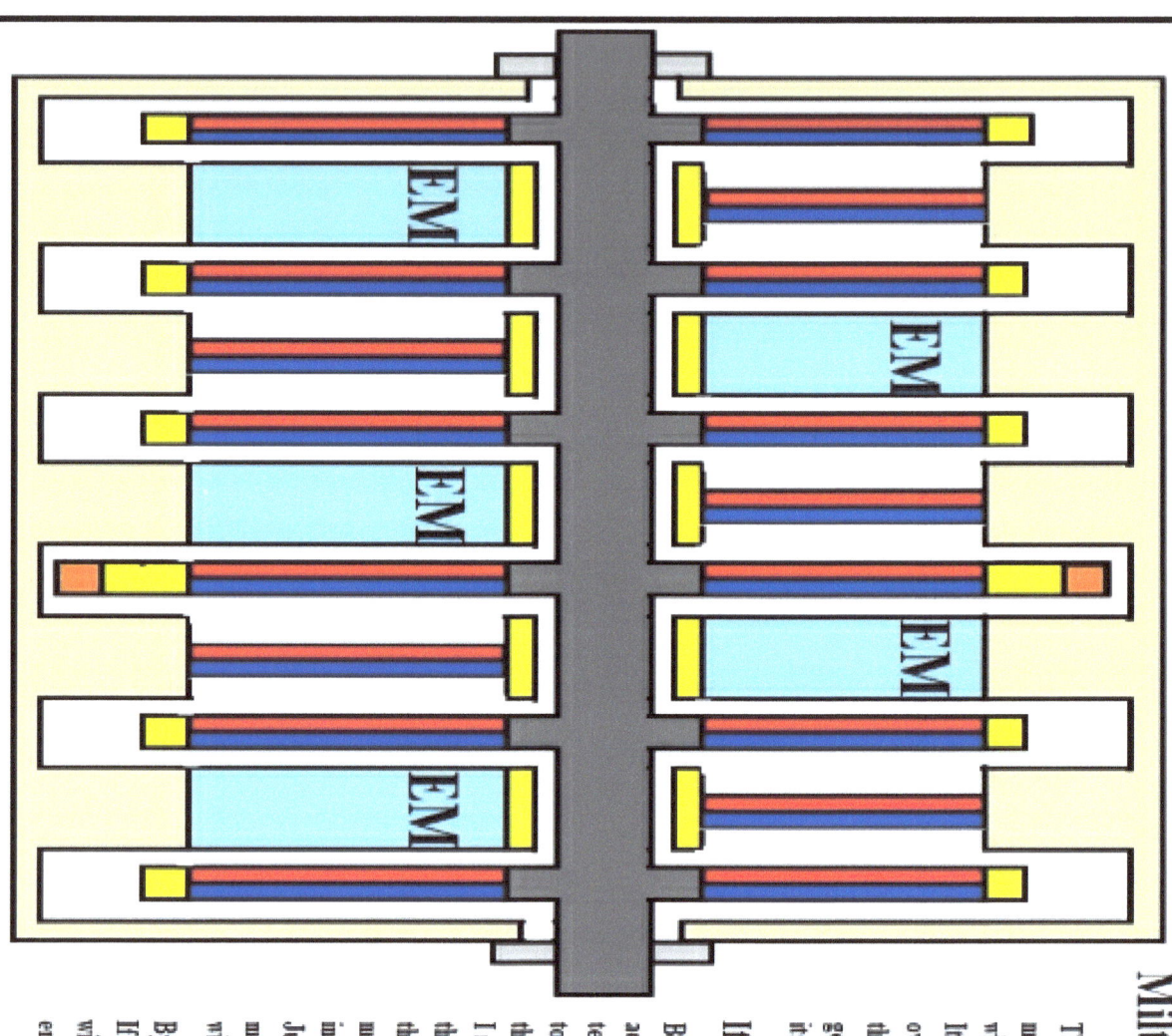

Milti-Disk Five to One PM to EM Disk Motor

The multi-disk format of the 5 to 1 PM to EM disk motor has the potential of moving motor technology from wagons being pulled by square wheels to using round wheels when power by the modified tank circuit with steering diodes.

It one one thing to have a motor that is overunity; but if it is bulky and is not much over unity, then does not make sence to use in in a vehicle. This motor design with the power circuit has more than enough power in torque to power an efficient generator to power the motor at the same time have enough power left over to have its shaft power the drive train of the vehicle.

If and only this is true about this new technology.

Being an old man soon to be living on social security; I will not be able to build and advance this technology. It is up to people like you who can see the potential of this technology; and then to build prototypes to prove its potential for the world. I have to admit that when I look at other peoples motor designs, it is hard for me to have that desire to build thier design because I always want to focus on my own designs. I reached out for help and I was asked to pay 18 times of what I was making for that help. I have not accepted money in the past and I am not seaking money in the future. What I am asking for is people to examine all the postings I have made at this site that includes much more details of this new technology which includes the modified tank circuit with steering diodes.

Just think of the good it would do for the average person of the world to have a motor that could provide all of the power they needed in a self contained system without having to buy fuel or electrical energy to operate it.

By building and posting the results of this new motor technology would be great. If the results are that the technology does not meet the OU status, I could live with that. Even if the results was a more efficient motor using less electrical energy per mile traveled in your car would be a big plus for everyone.

Jay Lunke modified 12-7-20

The Three Layer Electromagnetic Movement Technology can be used in linear applications as well.

Now if you can even get two or three times more range with this new motor design over magnetic motors, then the motor design will likely be worth it in the automotive industry. This is if it still has not reached over unity status. Now if it has over reached unity status at this point, then changing to these new motor designs is a no brainer.

Now the way this motor could function in an over unity vehicle is that the vehicle would have a battery like it does now and use the battery when you are driving the vehicle. Now when you park the vehicle, then the car would automatically go into the charge mode. It is the charge mode that will recharge the battery. This would be the best way to operate when using an over unity motor in the vehicle. The reason for this is because the motor would have to be too large for the car if it had to move the vehicle at the same time as to produce electrical energy. There is usually time each day to charge a car when not being driven. Now maybe in commercial vehicles you would want the larger motor to provide both the torque and charging of the battery at the same time.

I want to introduce you to the flow though motor. The flow through motor is a design I came up with about 40 years ago. The idea of the flow through motor was to optimize how close the permanent magnets would be to the electromagnets in order to maximize the strength of the motor. It was also the objective to maximize the size of the permanent magnet to the electromagnet.

I want to show the series to series four segment Three Layer Electromechanical movement because this technology is the closest to the technology used in the flow through motor upgrade. The drawings can only show the rotors on two side of the stator assembly. The flow through motors goal is to have rotor movement all around of the stator as possible in order to have as much motor strength as possible of the motor assembly. Notice the big improvement of these new designs over the old designs I wrote in my book named "A free gift that may over unity or free energy to the world".

Segment One Of Series Movement Of Technology

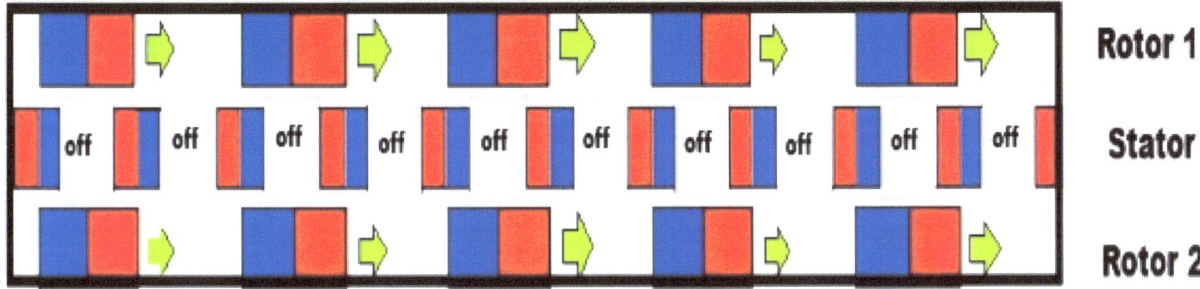

Beginning Of Movement

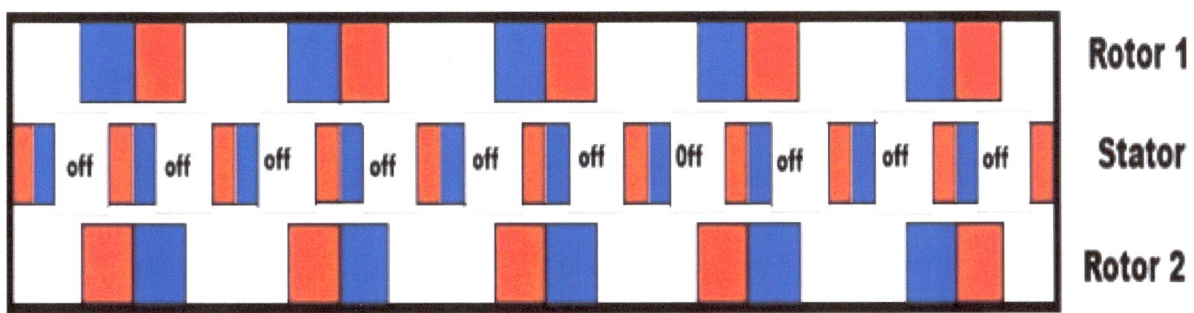

Finishing Movement

Jay Lunke   11-29-2025

The starting location of the first of a four segment travel is at the end of the 4th segment location. The momentum from the fourth segment travel will cause the rotor rotation to continnue to the right. The power is turned off for both electromagnet circuits. So the movement of the rotors is a result from the torque between the permanent magnets in both the stator and rotor assemblies causing posative rotation with no electrical energy being used in the motor assembly.

The rotor magnets are twice the size than the stator magnets in order to average out the length of the rotor magnets between the stator permanent magnets and the stator functional magnets that will be operating in the second and fourth segment configurations.

There are three Three Layer Electromechanical Movements that use the four segment movement: Series to series, parrellel to parellel and series to parrellel. This is one of them

Segment Two Of Series Movement Of Technology

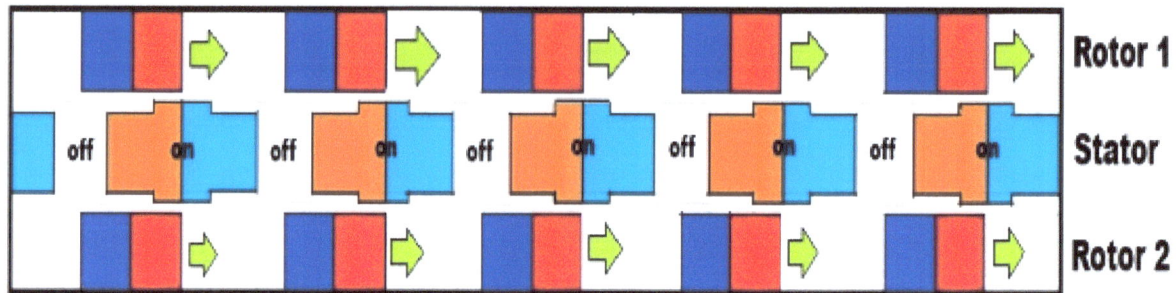

Beginning Of Movement

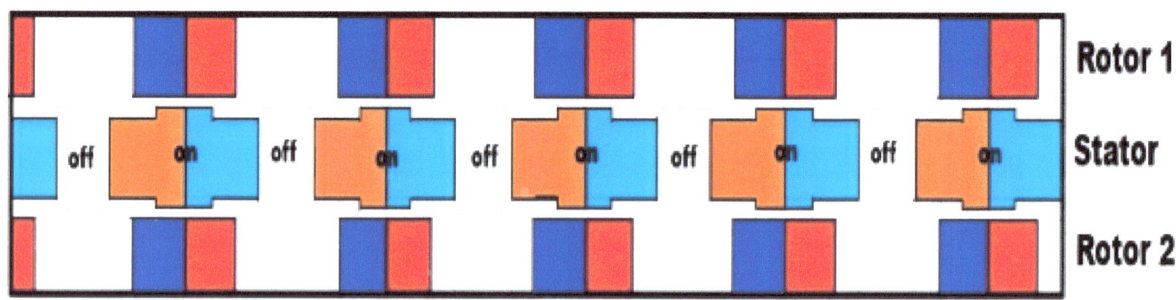

Finishing Movement

In the second segment of travel for the Three Layer Electro-mechanical Movement, every other electromagnet in the stator assembly is turned on. The other electromagnets have the power off. Each electromagnet with the power on along with the two permanent magnets adjacent to them create a functional magnet. They are shown by the orange and light blue colors. This reconfigures the stator assembly by reducing the magnets in the stator in half. The poles in the stator assembly are changed because of this. This segment has torque from two permanent magnets and one functional magnet.

Jay Lunke 11-29-2029

Segment three Of Series Movement Of Technology

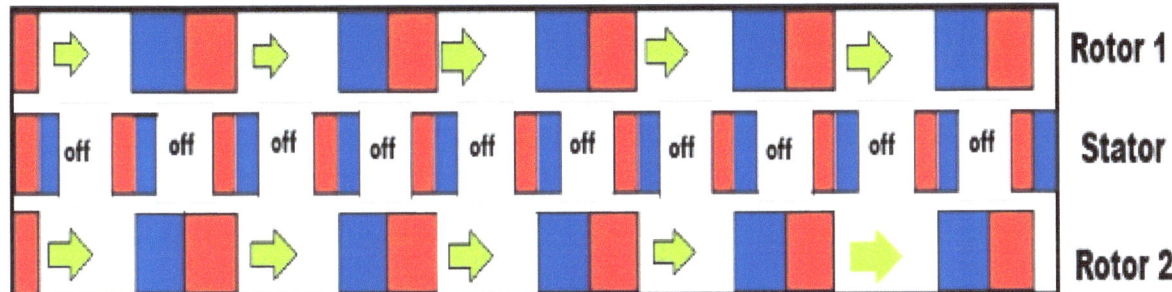

Beginning Of Movement

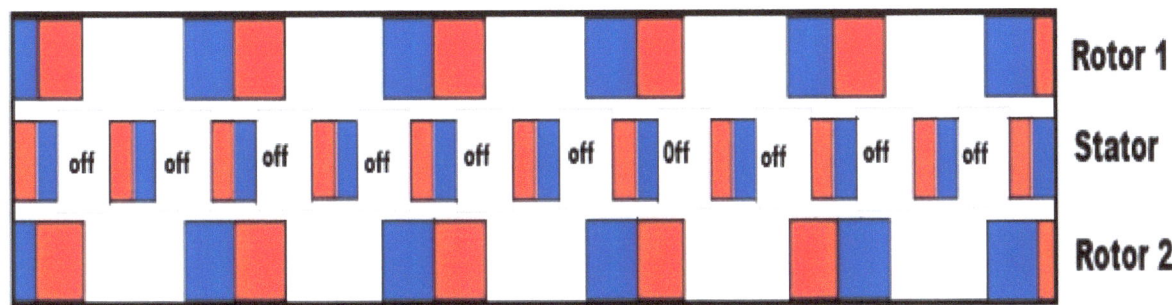

Finishing Movement

The third segment of travel is very simular to the first segment of travel. The movement of the two rotors in this segment does not require any electrical energy in order for the movement to occur. This is because at the end of the second segment movement, the power to all the electromagnets is turned off. This creates a torque between the permanent magnets between the two motor assemblies and the stator assembly. It is this torque that causes the movement of the rotors in this segment of travel without electrical energy.

This series to series Three Layer Electromechanical Movement design is ideal for upgrading one of my first motor designs called flow through motor technology Because that technology needed power all the time to the motor but now only needs power for 50% of the time and deviding that in half for each EM circuit.

Jay Lunke   11-29-2025

Segment Four Of Series Movement Of Technology

Beginning Of Movement

Finishing Movement

The fourth segment is simular to segment two. The difference is that the electromagnets that were off in segment two are now turned on in this segment. The electromagnets that ere turned on in segment two are now turned off in this segment. The torque from the two rotors and the turque from the functional magnet in the stator assembly are opderating the motor. All four segments of travel have forward torque on them. When looking at all the torques from all four segments of travel, you find that you get torque from 10 permanent magnets and two functional magnets. This make a ratio of five to one. The torque ratio of a magnetic motor used in most electric cars today is at a ratio of one to one. This technology in theory should be used over the magnetic motors used today.

Jay Lunke 11-29-2025

I have updated the flow through motor design to that of the Three Layer Electromechanical Movement Technology. I have done this with two different designs. The first drawing shows the electromagnets on the stator assembly. The second design shows the electromagnets on the rotor assembly. Each of these designs have their plusses and minuses. Now the second design has better efficiency because the electromagnets use less wire in them because the size requirement is smaller. Now the number of permanent magnets to electromagnets is 3 to 1, the performance could be greater than the 5 to 1 PM to EM motor designs I have made. The reason for this is because the permanent magnets in the stator assembly of the second design, wrap themselves around most of the electromagnet. This maximizes permanent magnet torque coverage to electromagnet coverage. The more push you get between the rotor and stator permanent magnets means a stronger free push in the motor design. Which design is better can only be determined by building both motors and comparing the performance of them? So how about building them.

Notice the difference in size of rotor magnets to stator magnets. The big thing to know is that the magnet poles have the controls of the torque in the motor. So, follow the poles to have a good understanding of what is happening to the torque in the motor's movement.

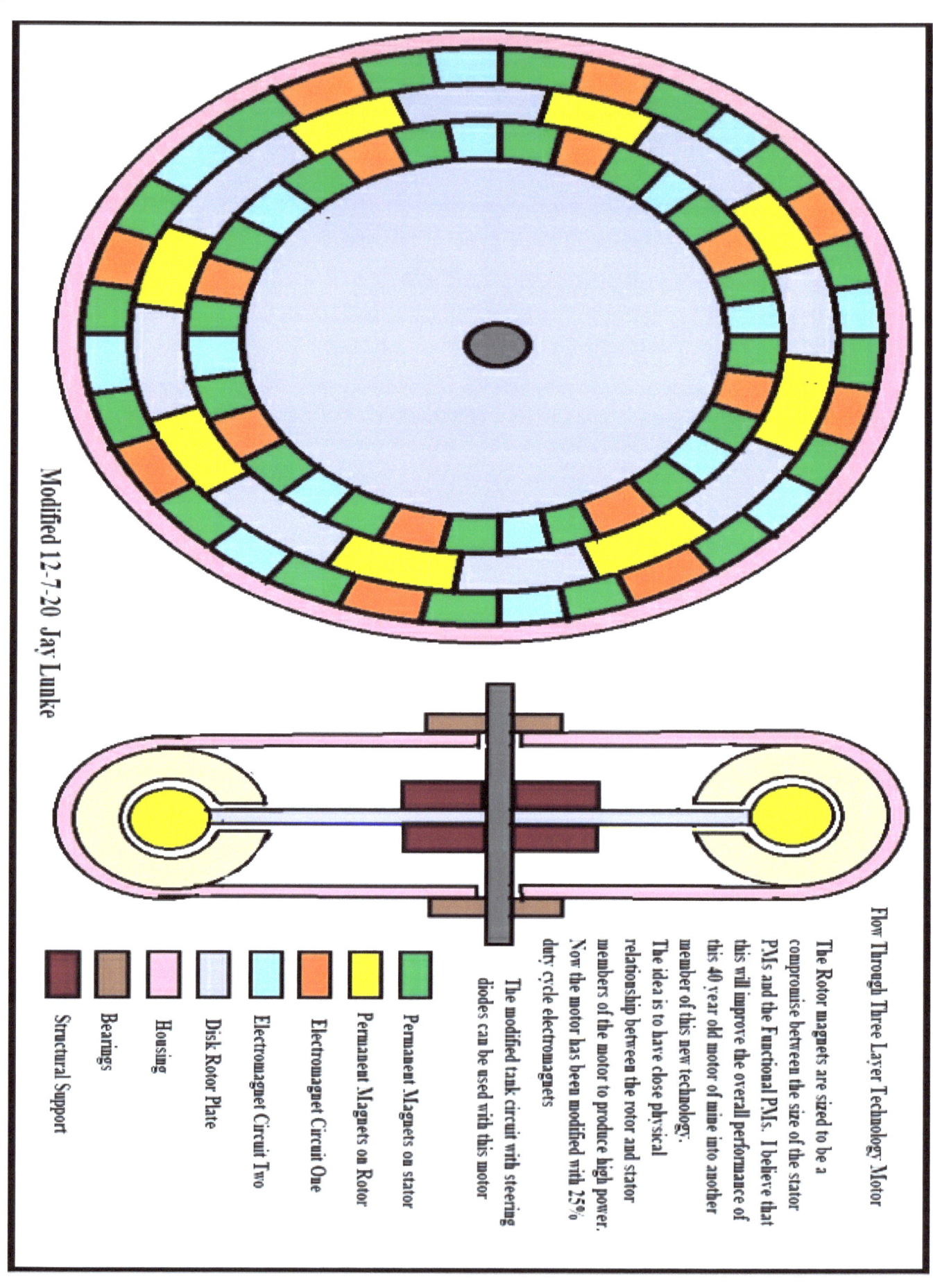

PERMANENT MAGNET TORQUE HARVESTING

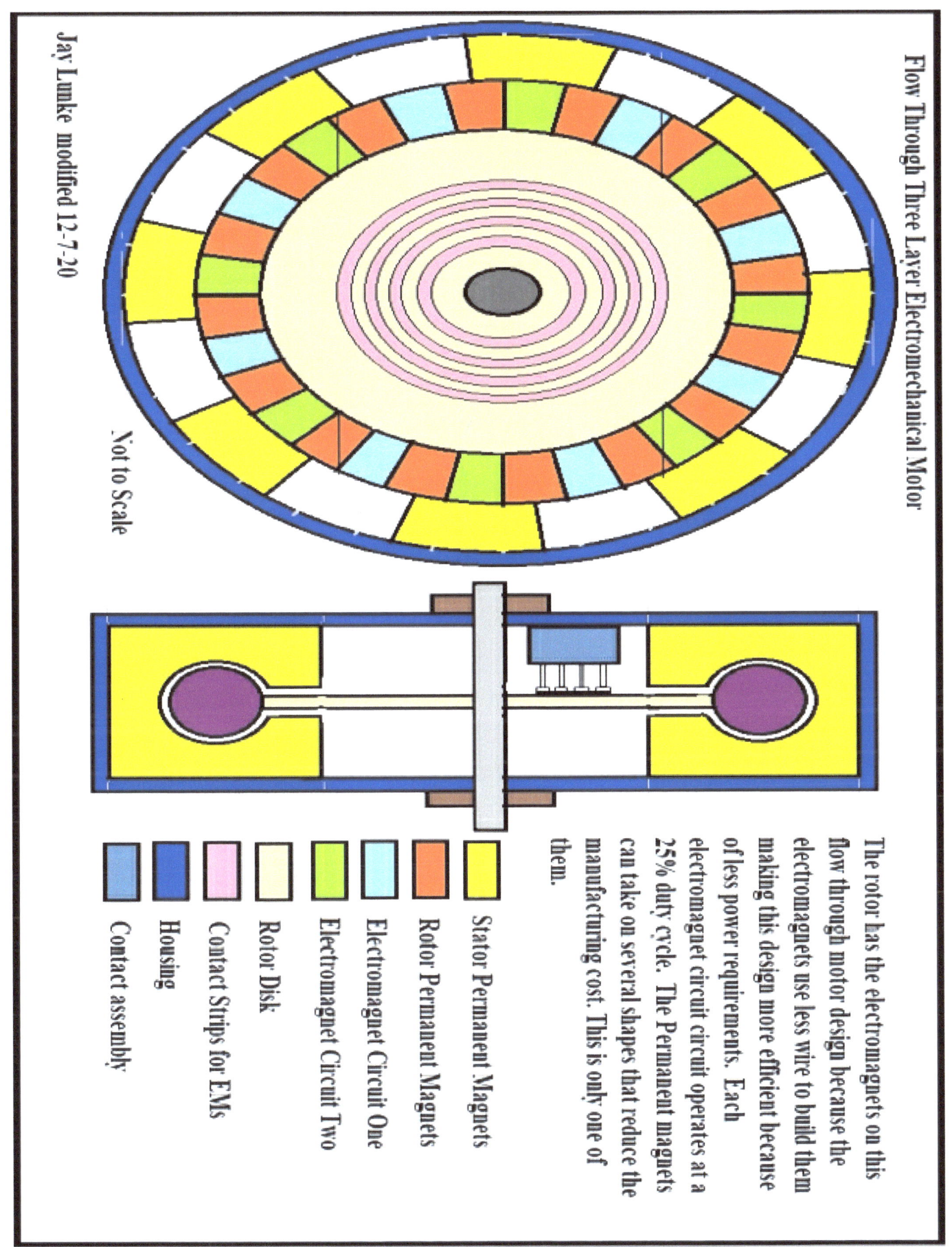

Flow Through Three Layer Electromechanical Motor

Jay Lumke modified 12-7-20

Not to Scale

The rotor has the electromagnets on this flow through motor design because the electromagnets use less wire to build them making this design more efficient because of less power requirements. Each electromagnet circuit operates at a 25% duty cycle. The Permanent magnets can take on several shapes that reduce the manufacturing cost. This is only one of them.

- Stator Permanent Magnets
- Rotor Permanent Magnets
- Electromagnet Circuit One
- Electromagnet Circuit Two
- Rotor Disk
- Contact Strips for EMs
- Housing
- Contact assembly

As you examine the drawings, you can see how close the rotor is to the stator magnets at all times through the operation of the motor. This means a lot of torque between the rotor and stator assemblies. The biggest concern I have is that the permanent magnets in the rotor assembly will have so much torque applied to them, this may reduce the strength of the rotor magnets over time. Now some of the new rare earth magnets may be able to hold up to these forces.

Now the following drawing shows a linear motor. This motor uses the 5:1 PM to EM torque ratio in it's operation. This linear design takes what I have in the disk motor and stretches it out into a straight line. It also takes advantage of the modified tank circuit with stearing diodes which will be discussed later on in this book. There are so many linear movements that this technology can be used in. This is only one of them.

Linear technology will be around awhile because like a train, you can move a lot of payload with a few operators. As technologies increase, human expence will become more of a thing companies will try to reduce in order to be more compeditive in the world.

A free energy linear power will have unlimited applications.

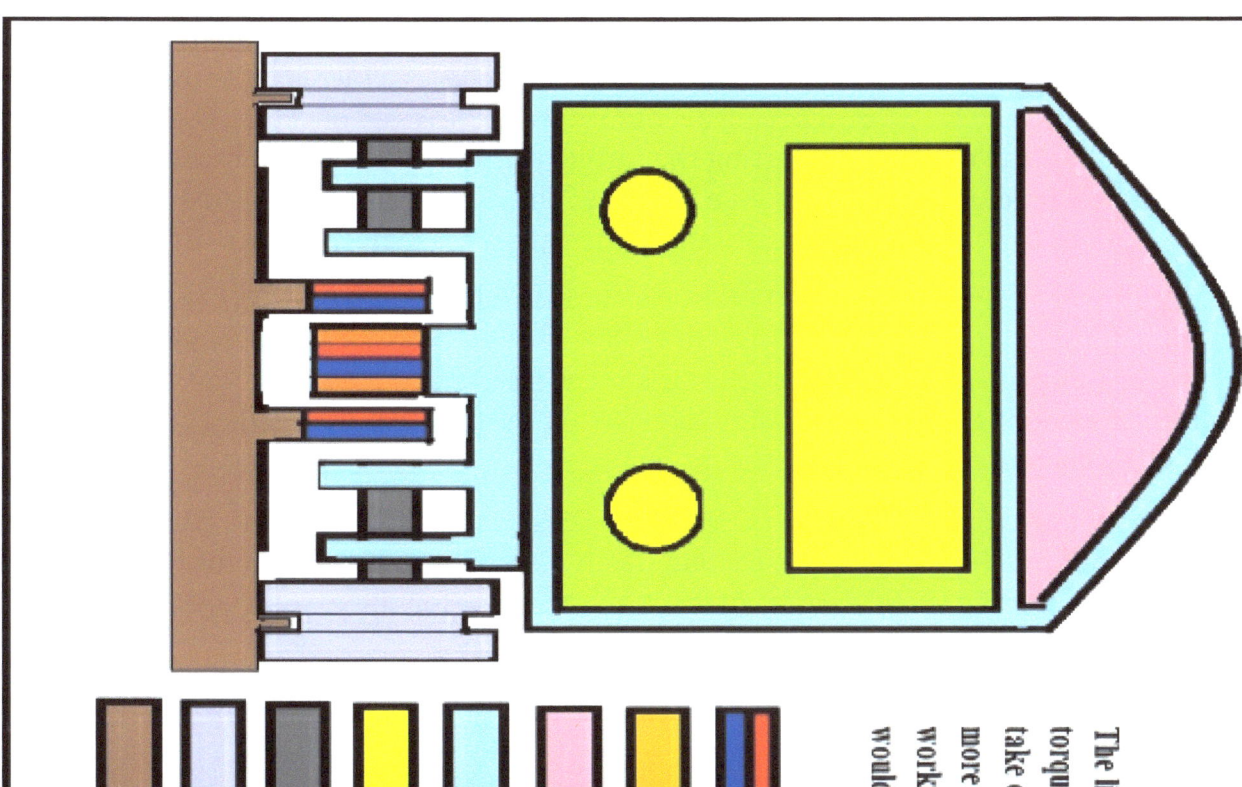

Mass Transit Linear Motor

The linear motor is great for mass transit transportation. With the 5:1 PM to EM torque ratio and the modified tank circuit with steering diodes, this new technology could take over the rail system, the sub-way system for a start. Now this system will be a lot more efficient than todays mass transit systems. And should these two technologies working together in this system, outside energy may not be required at all. These savings would be passed down to the customers.

The linear motor operates with the same four segments of travel that the circular motor movement has. The stator is connected to the vehicle and the two rotors are connected to the track. Now this drawing does not show a generator on it. So with this design, the motor will need a power source. But the efficiency of this mass transit system will be so efficient, that it would be worth using this design over current designs.

Now the wheels could be replaced with magnetic bearings, but that is a lot of expense unless it is the track on amusement park.

- Permanent Magnet
- Electromagnets
- Equipment Bay
- Structure
- Glass
- Shaft
- Tires
- Track

Modified 12-7-20 Jay Lunke

Now to farther improve the 5 to 1 PM to EM ratio motors over current magnetic motor designs then the incorporation of the modified tank circuit with steering diodes can increase the motor efficiency increasing the range of travel and or reaching over unity status. Instead of saying over unity status, it is better to say that the motor is harvesting the torque from the permanent magnets in order to perform work. The following diagram is an example of the circuit that can be used to accomplish that. This circuit can be used in other people's motor designs with some minor modifications.

The tank circuit was designed about 100 years ago and is used in many circuits today. It is used mostly in transmitteres and receivers because they work very well when at the resonant point. I have modified the tank circuit so that one leg of the tank is the electromagnet circuit and the other leg is the capacitor. The steering diodes is a modification so that the tank circuit does not have to be at the resonant point to operate efficiently with the motor.

A tank circuit is highly efficient because the energy cycles back and forth in the circuit. The tank circuit is topped off in order to allow it to sustain itself. I mimic this activity in the circuit in order to keep the energy used by the electromagnets at a minimum for the system response to power consumption.

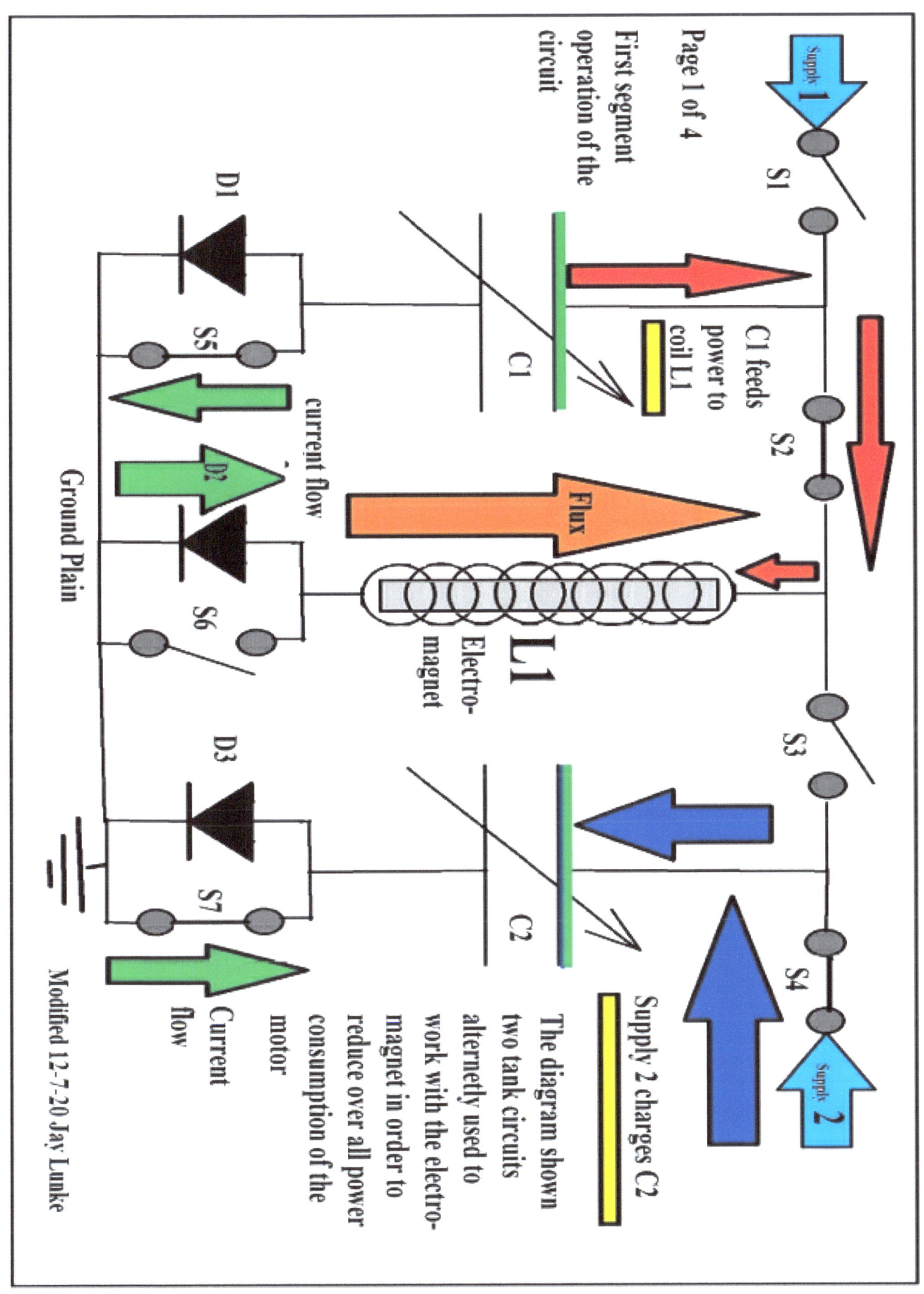

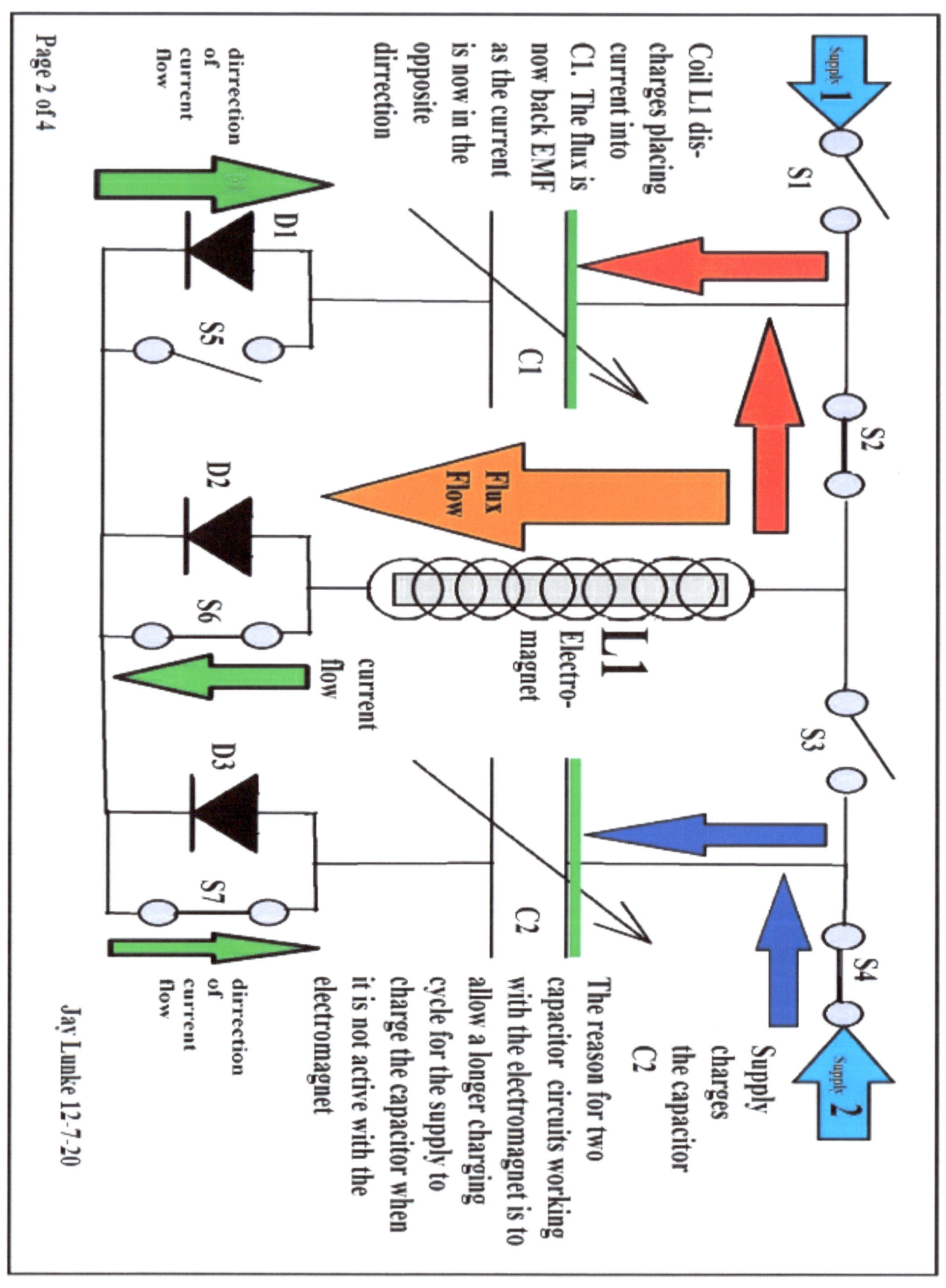

PERMANENT MAGNET TORQUE HARVESTING

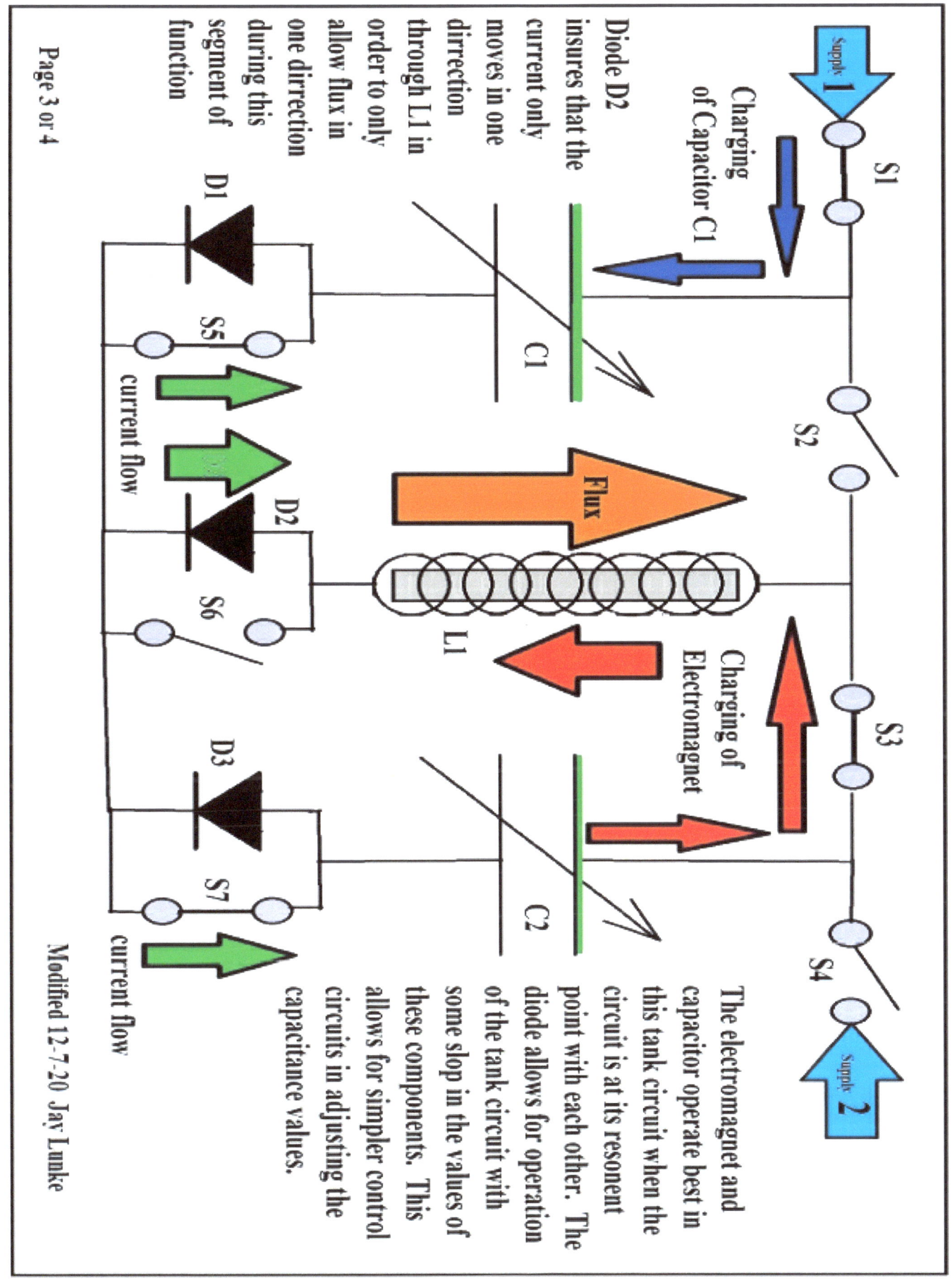

Charging of Capacitor C1

Diode D2 insures that the current only moves in one dirrection through L1 in order to only allow flux in one dirrection during this segment of function

Charging of Electromagnet

The electromagnet and capacitor operate best in this tank circuit when the circuit is at its resonent point with each other. The diode allows for operation of the tank circuit with some slop in the values of these components. This allows for simpler control circuits in adjusting the capacitance values.

Modified 12-7-20 Jay Lunke

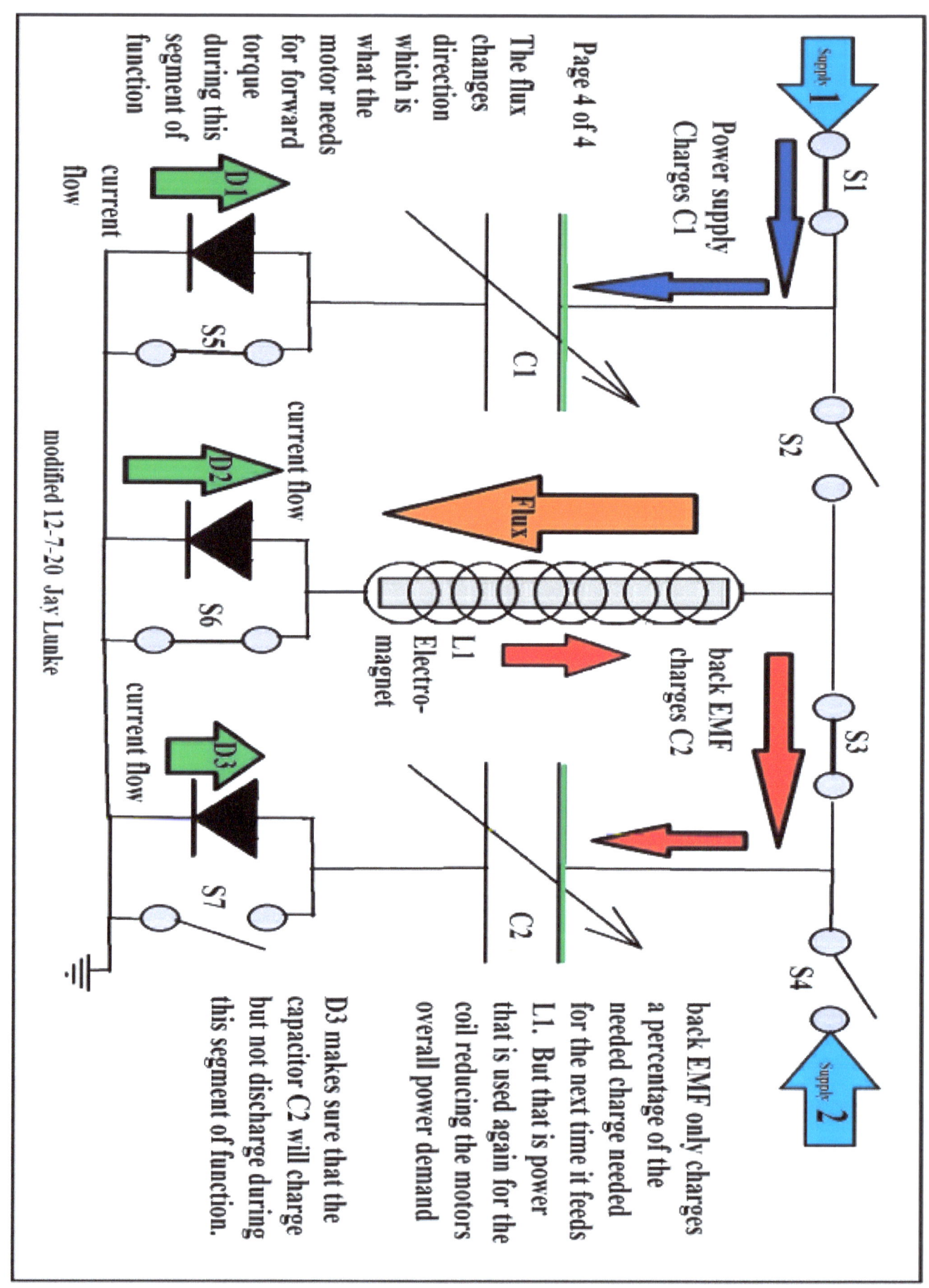

Since each of the two electromagnets circuits operate at a 25% duty cycle, then this circuit is ideal to reduce the energy usage of the motor. Most Three Layer Electromechanical Movement Motor Designs can use this power circuitry in them.

Now this tank circuit can be used for other motors in order to capture and reuse a lot of the electrical energy that is used to build the magnetic flux in the electromagnets as well. The four drawing above are set up to show you what is happening in each segment of travel when used to drive a motor.

The next drawing shows how one of my rotor motors can use the tank circuit to operate the motor. The drawing is triggered by reed circuits. Other options like optical signals being used for switching may be better options to signal the switching. Higher speeds require something faster than reed switches. The main concept is to design a resonant circuit to create an efficient motor drive circuit as possible.

Defining logic in tank circuits used in three layer disk motor

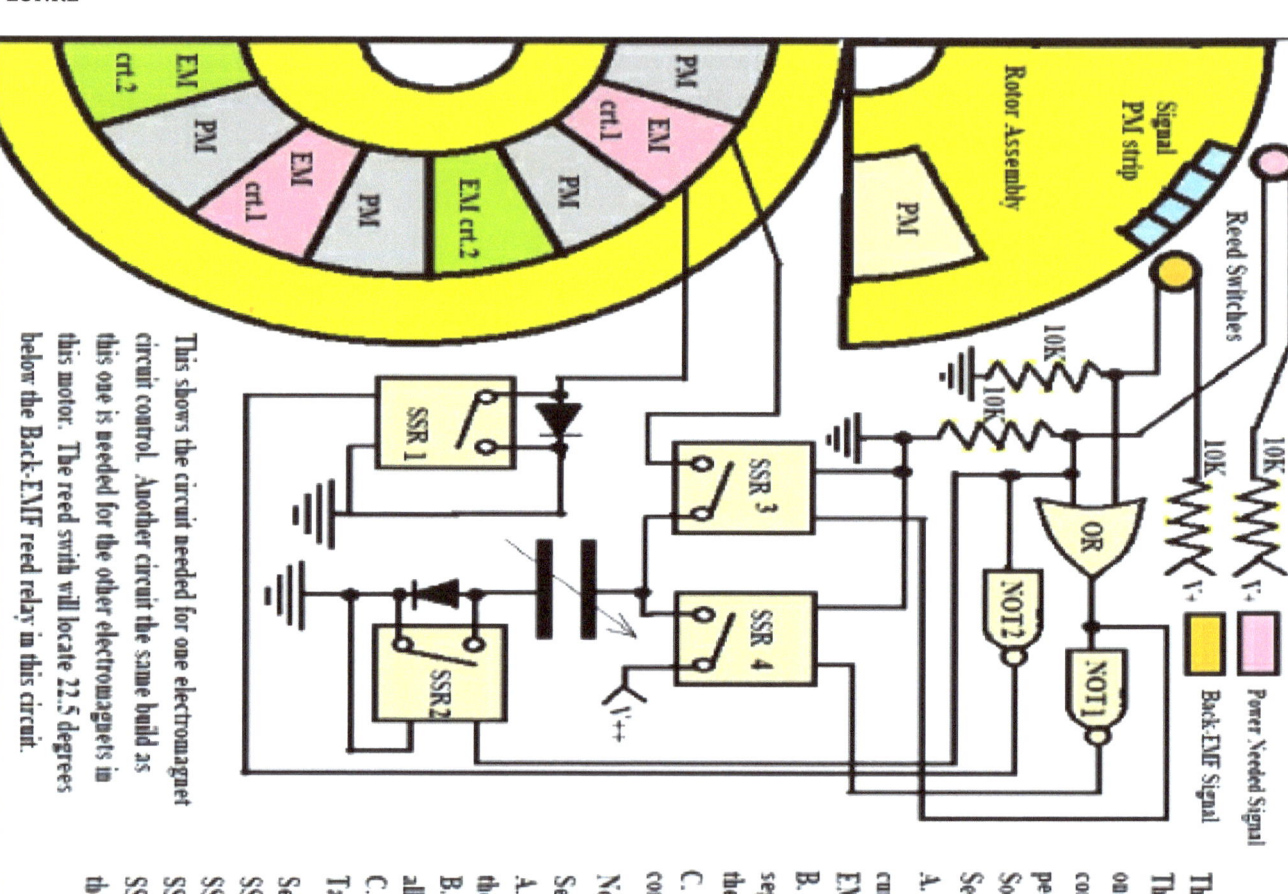

The motor is devrided into 16 segments of travel each having 22.5 degrees to thier travel. The stator has two electromagnet circuits, one to operate in the first segment of travel and the other one to operate in the third segment of travel. During the second and forth segments of travel, the air core electromagnets are turned off and the permanent magnet in the rotor assembly interact with the permanent magnets in the stator assemblies to rotate the rotor 22.5 degrees.

So the following logic occures in the circuit as follows;

Segment 1: The Power Needed Signal reed switch closes creating a posative signal on its signal line

A. The signal goes through NOT2 being a low signal. This opens the switch in SSR1 restricting the current flow in one direction. As the current flows from the Tank capacitor, we do not want back EMF to occur during this time.
B. The signal is feed to an OR gate. The OR gate will create a posative signal that lasts through segments 1 and 2. This will tie the electromagnet and capacitor together as a Tank circuits through these to segments of travel using SSR3.
C. The OR gate also feeds NOT1 gate that goes to SSR4 to make sure the the power supply is not connected to the Tank capacitor at this time.

Note: SSR2 stays in the same condition. That of shorting the diode so current flows into the coil.

Segment 2: The Back-EMF signal closes creating a posative signal on its signal line.
A. This Goes to the OR gate which feeds SSR3 to keep the Tank coil and capacitore connected so the back-EMF can be collected into the capacitor.
B. The signal goes to the SSR2 relay to open the switch activating the function of the diode. This allows charging only for the capacitor in the Tank circuit.
C. Since the "Power Needed Signal" is now off, SSR1 is shorting out the diode at the bottom of the Tank electromagnet allowing the current to flow into the capacitor.

Segment 3 and 4 are the same:
SSR1 Through NOT2 causes swtich to close shorting diode.
SSR2 is open so the diode at the bottom of the Tank circuit is active.
SSR3 has a low signal from the OR gate causing the electromagnet circuit to be open.
SSR4 comes from a NOT OR logic causing the switch to be closed in order to top off the charge of the Tank capacitor.

This shows the circuit needed for one electromagnet circuit control. Another circuit the same build as this one is needed for the other electromagnets in this motor. The reed swith will locate 22.5 degrees below the Back-EMF reed relay in this circuit.

Modified 12-7-20 Jay Lunke

The following drawing describes the modified tank circuit with steering diodes operating an electromagnet of a motor assembly. It would not be hard for a motor designer to modify this circuit for motor designs that have other duty cycle requirements.

Since diodes have a voltage drop that reduces the efficiency of the circuit, there other options that can be designed into the circuit that will make the circuit even more efficient than it is now.

The current drawing is at its simplest form that shows you the concept of building a circuit that simulates a resonate circuit with a motors drive coil.

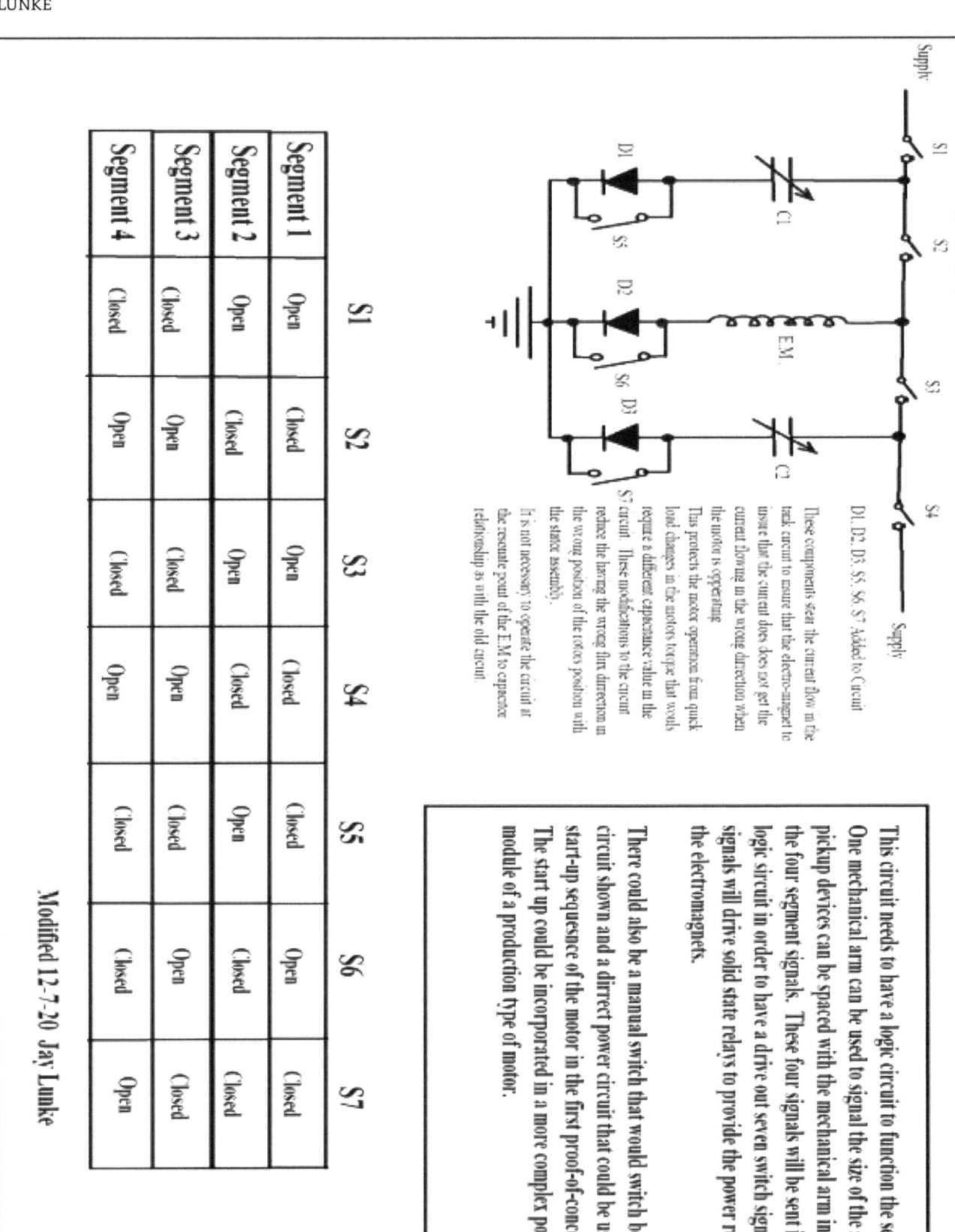

D1, D2, D3, S5, S6, S7 Added to Circuit

These components seat the current flow in the task circuit to ensure that the electro-magnet to unsure that the current does does not get the current flowing in the wrong direction when the motor is operating. This protects the motor operation from quick load changes in the motors torque that would require a different capacitance value in the S7 circuit. These modifications to the circuit reduce the having the wrong flux direction in the wrong position of the rotors position with the stator assembly.

It is not necessary to operate the circuit at the resonate point of the E.M. to capacitor relationship as with the old circuit.

This circuit needs to have a logic circuit to function the seven switches. One mechanical arm can be used to signal the size of the segment. Four pickup devices can be spaced with the mechanical arm in order to have the four segment signals. These four signals will be sent into a simple logic circuit in order to have a drive out seven switch signals. These signals will drive solid state relays to provide the power requirements of the electromagnets.

There could also be a manual switch that would switch between this circuit shown and a direct power circuit that could be used for the start-up sequence of the motor in the first proof-of-concept motor. The start up could be incorporated in a more complex power/control module of a production type of motor.

	S1	S2	S3	S4	S5	S6	S7
Segment 1	Open	Closed	Open	Closed	Open	Open	Closed
Segment 2	Open	Closed	Open	Closed	Open	Closed	Closed
Segment 3	Closed	Open	Closed	Open	Closed	Open	Closed
Segment 4	Closed	Open	Closed	Open	Closed	Closed	Open

Modified 12-7-20 Jay Lunke

The following drawing is one example of a variable capacitor that can be used in the tank circuits.

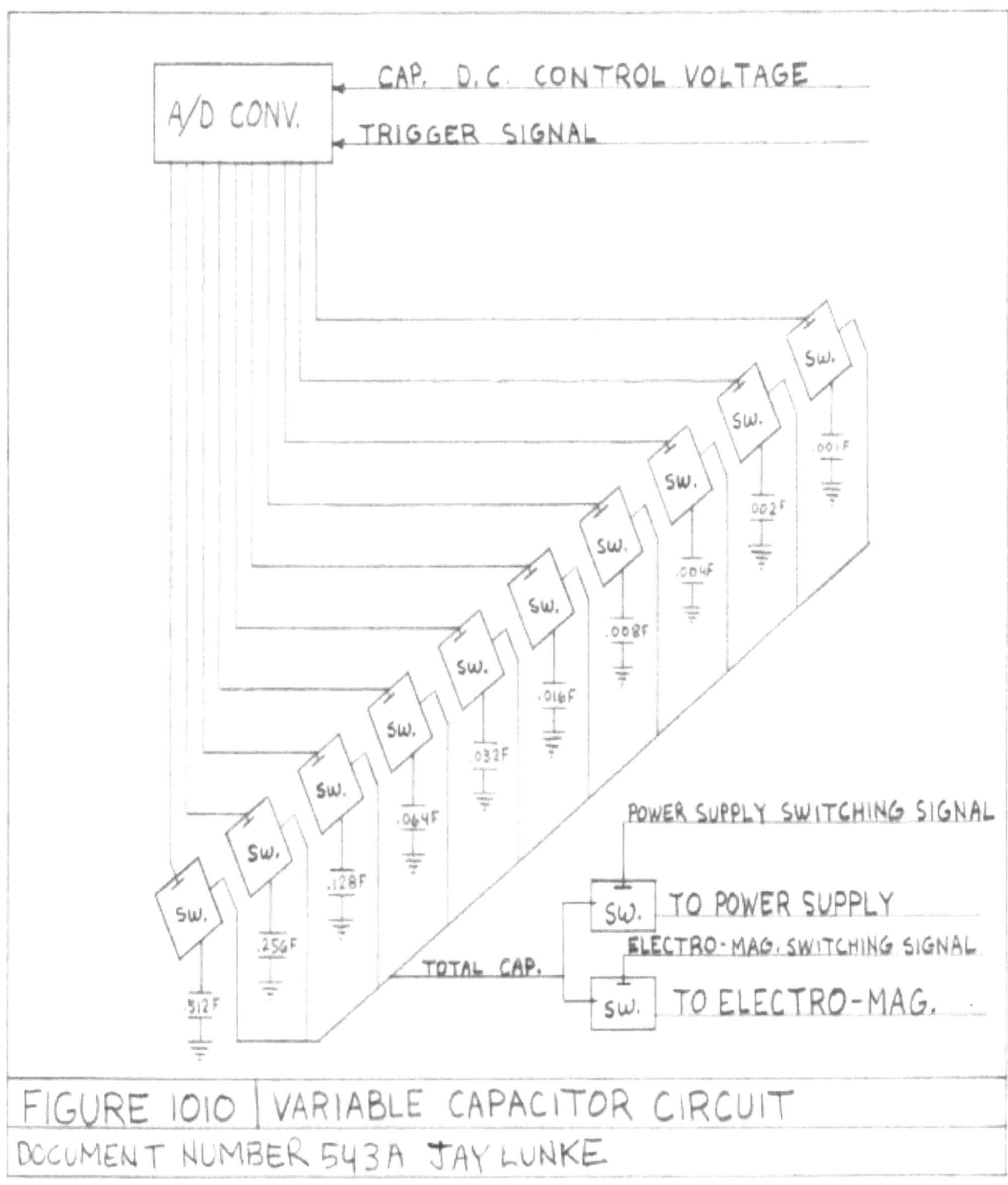

FIGURE 1010 | VARIABLE CAPACITOR CIRCUIT
DOCUMENT NUMBER 543A JAY LUNKE

Now if you want to have a very simple power control of the motors using this new technology, then the following diagram is a good design to do this.

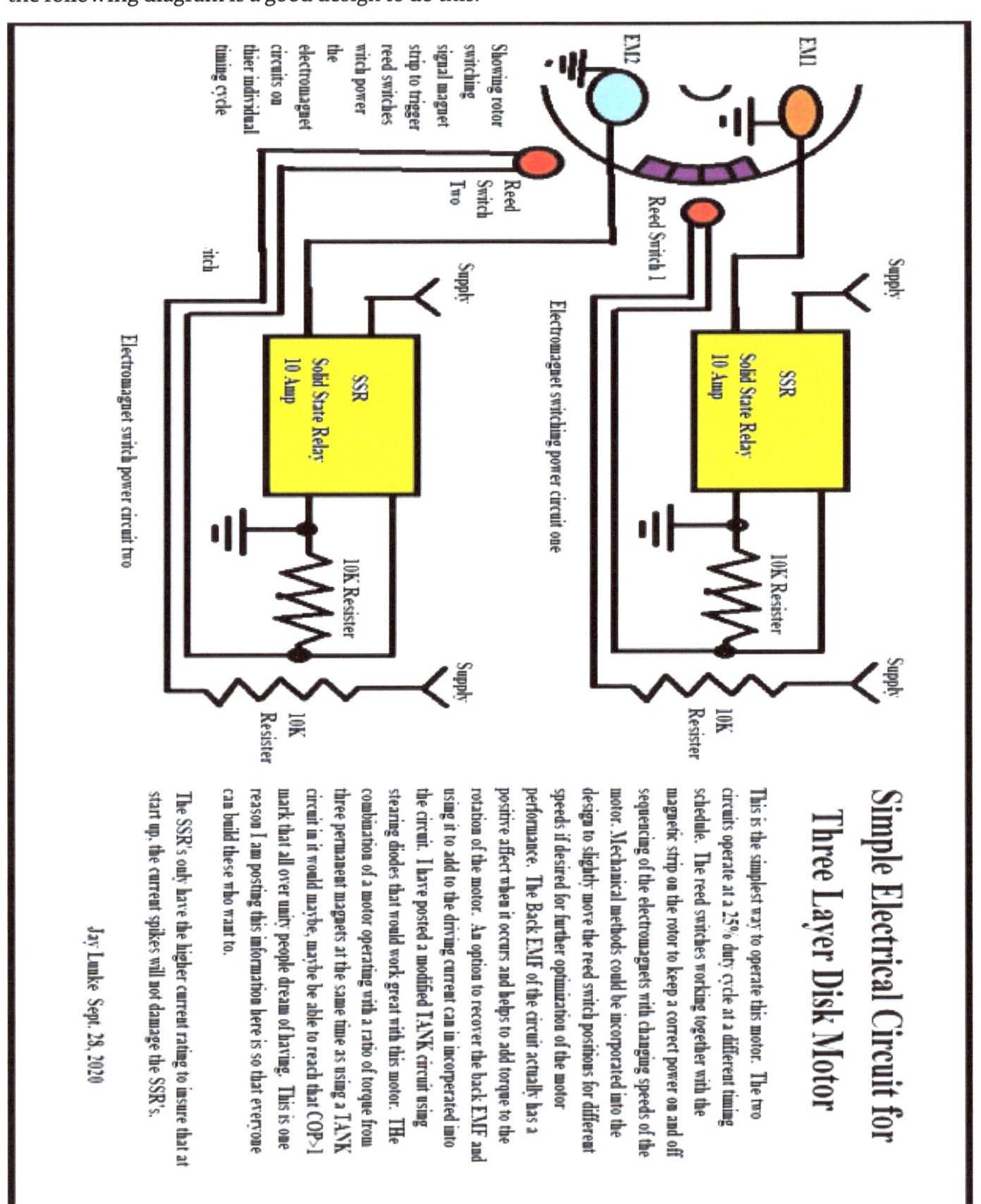

Now the Three Layer Electromechanical Movement Technology can operate with the other orientations of the rotor permanent magnets in relationship to the stator electromagnets. The next four drawings describe one of these motor options in four segments of rotation travel. These four segments repeat themselves over and over again.

Now these drawings show a poor design in magnet spacing. I use this size in order to show you in an easy way what the theory is in the movements. In real life, the magnets interact with each other at the poles, so in the drawings, the functional magnets become so big that there is little interaction toward the middle of the functional magnet with respect to the permanent magnet. So, what would need to be done is to place several more segments of travel into the circle so that the functional magnets become much smaller in size. This smaller size produces more interaction between the rotor and stator assembly.

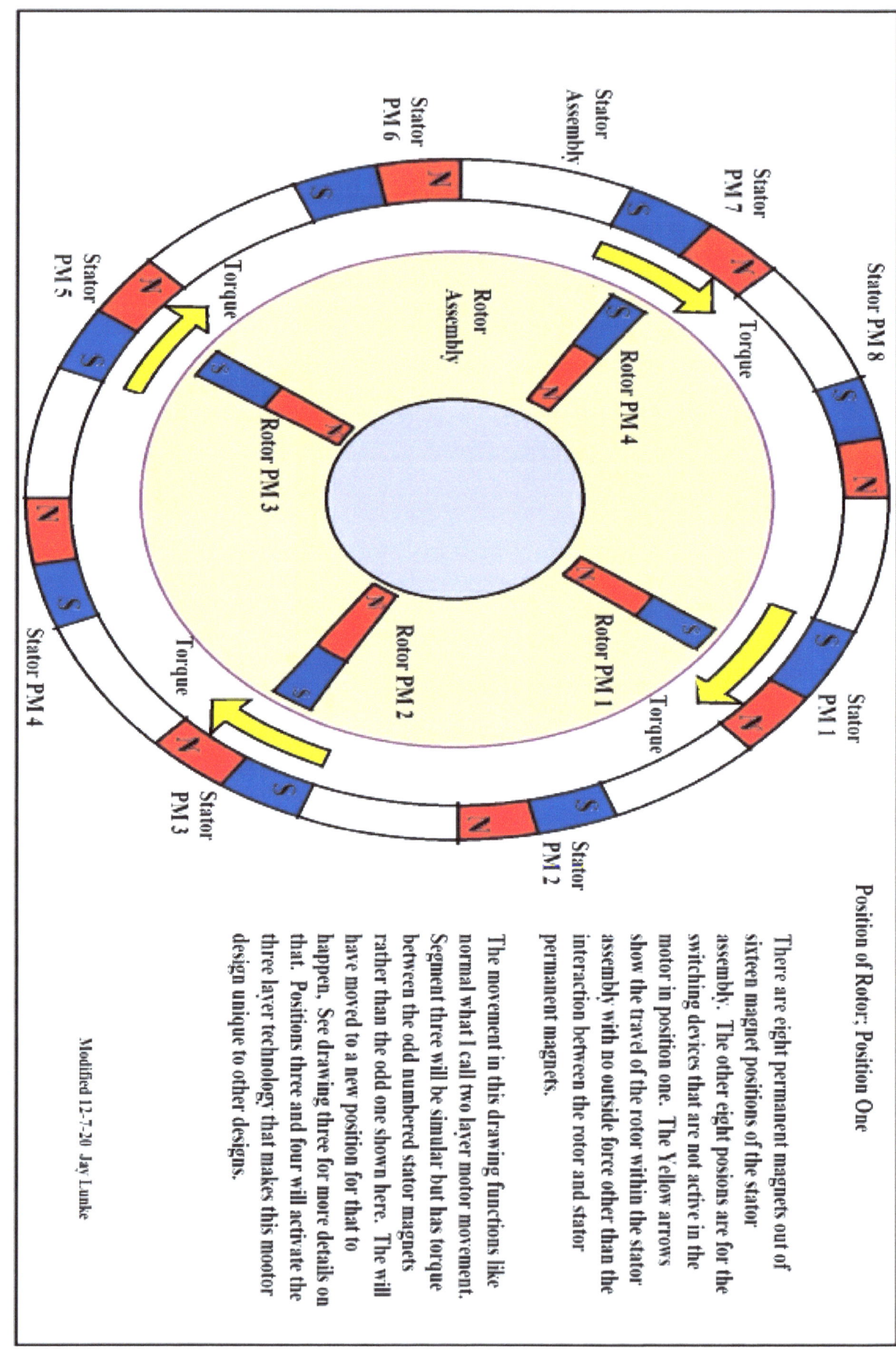

Position of Rotor: Position One

There are eight permanent magnets out of sixteen magnet positions of the stator assembly. The other eight posions are for the switching devices that are not active in the motor in position one. The Yellow arrows show the travel of the rotor within the stator assembly with no outside force other than the interaction between the rotor and stator permanent magnets.

The movement in this drawing functions like normal what I call two layer motor movement. Segment three will be simular but has torque between the odd numbered stator magnets rather than the odd one shown here. The will have moved to a new position for that to happen. See drawing three for more details on that. Positions three and four will activate the three layer technology that makes this mootor design unique to other designs.

Modified 12-7-20 Jay Lunke

PERMANENT MAGNET TORQUE HARVESTING

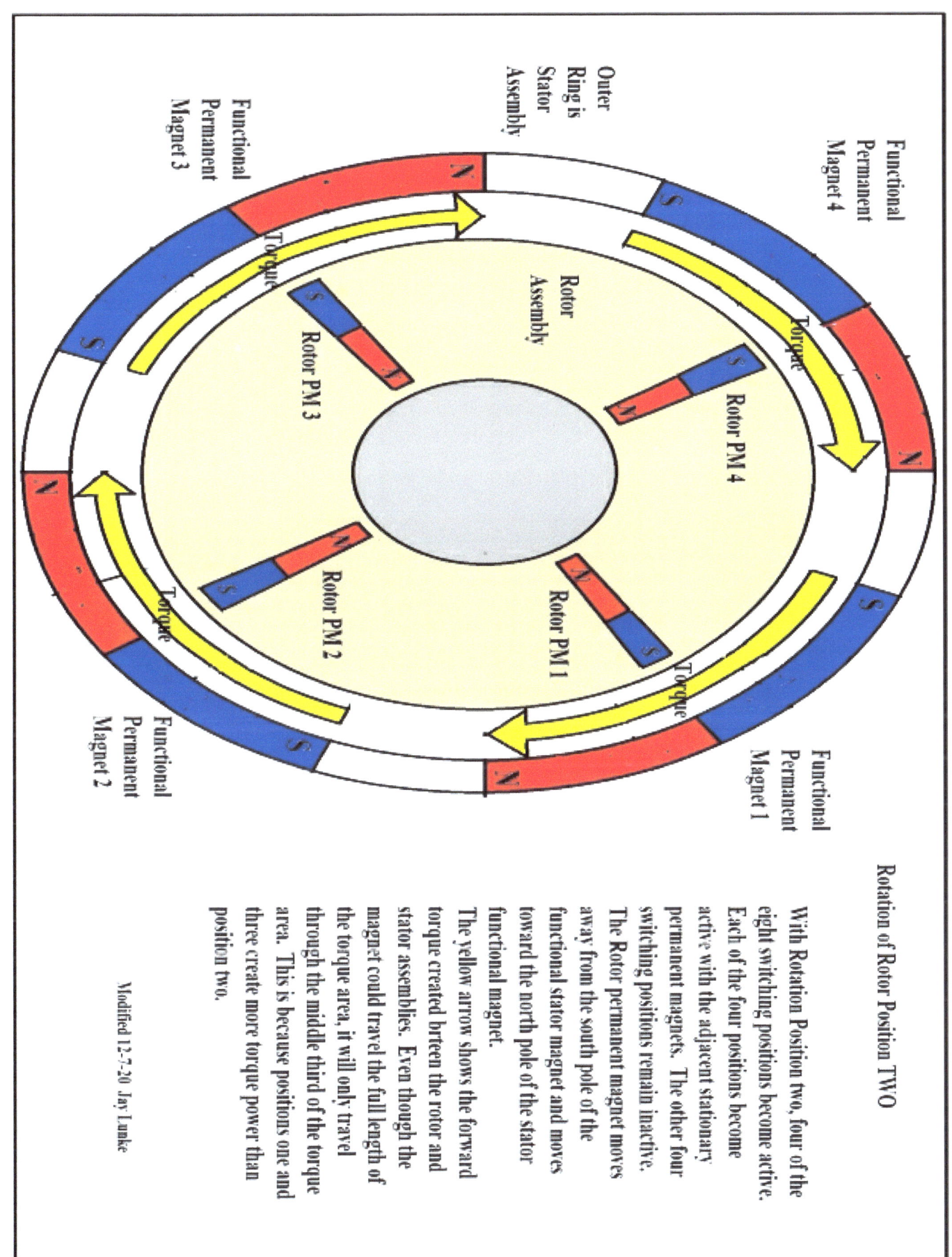

Rotation of Rotor Position TWO

With Rotation Position two, four of the eight switching positions become active. Each of the four positions become active with the adjacent stationary permanent magnets. The other four switching positions remain inactive. The Rotor permanent magnet moves away from the south pole of the functional stator magnet and moves toward the north pole of the stator functional magnet.

The yellow arrow shows the forward torque created brteen the rotor and stator assemblies. Even though the magnet could travel the full length of the torque area, it will only travel through the middle third of the torque area. This is because positions one and three create more torque power than position two.

Modified 12-7-20 Jay Lunke

55

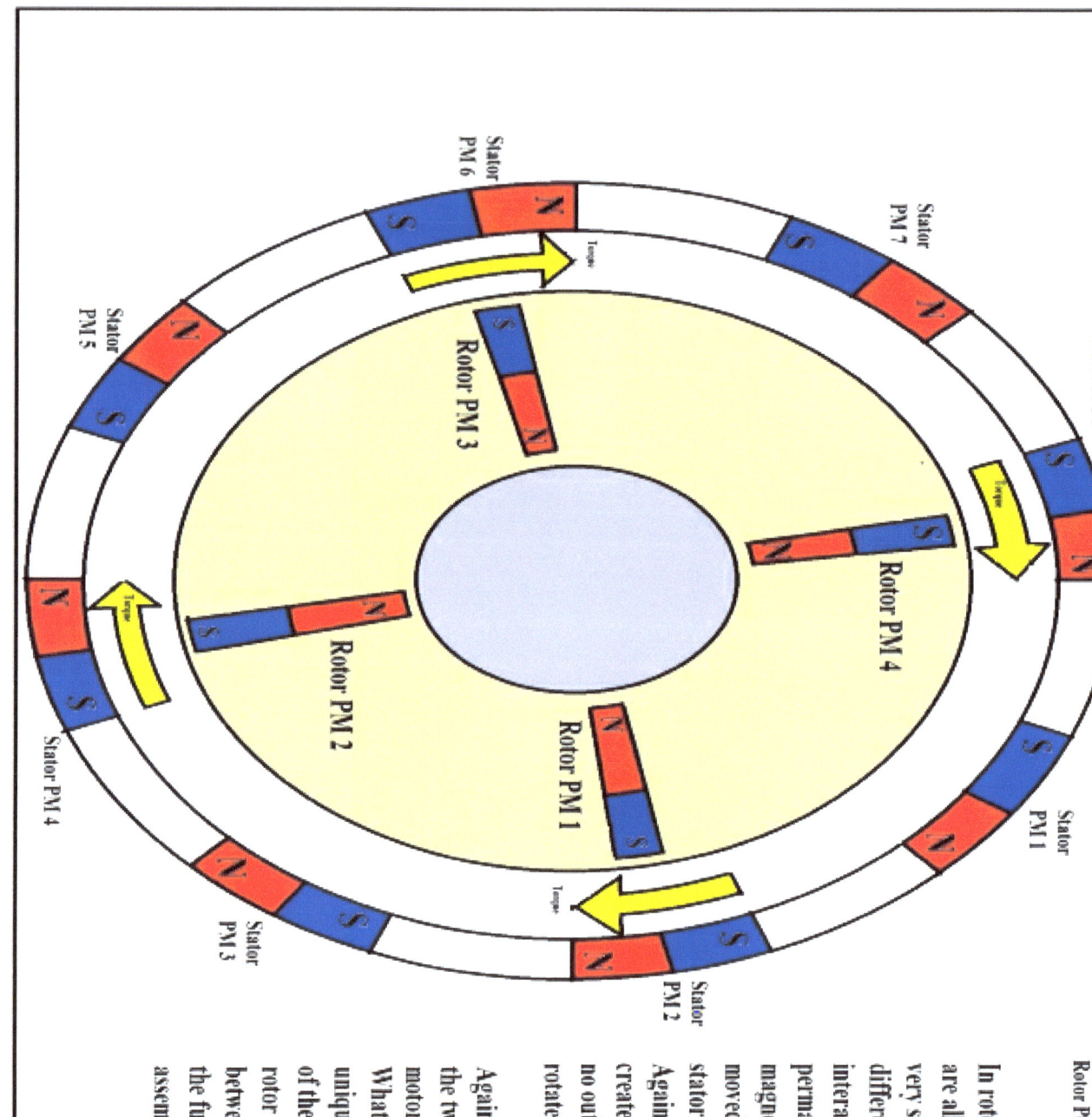

Rotor Position: Position three

In rotor travel position three the switching PMs are all inactive. The function in position three is very simular to that of position one. The big difference is that the rotor magnets are interacting with the even numbered stator permanent magnets instead of the odd numbered magnets. The reason is that the rotor has now moved to a new position in relationship to the stator assembly.

Again the yellow arrows show the torque that creates the rotor movement in the motor. Again no outside energy is needed in this position to rotate the rotor in the motor.

Again this position is operating in what I call the two layer format that is simular to other motors designs that can be found on the WEB. What makes the three layer design technology unique to others is the constant re-configuring of the stator assembly during the rotation of rotor providing constant forward torque between the rotor and stator assembly during the full rotation of the rotor in the stator assembly.

Jay Lunke modified 12-7-20

PERMANENT MAGNET TORQUE HARVESTING

Jay Lunke   June 1, 2020

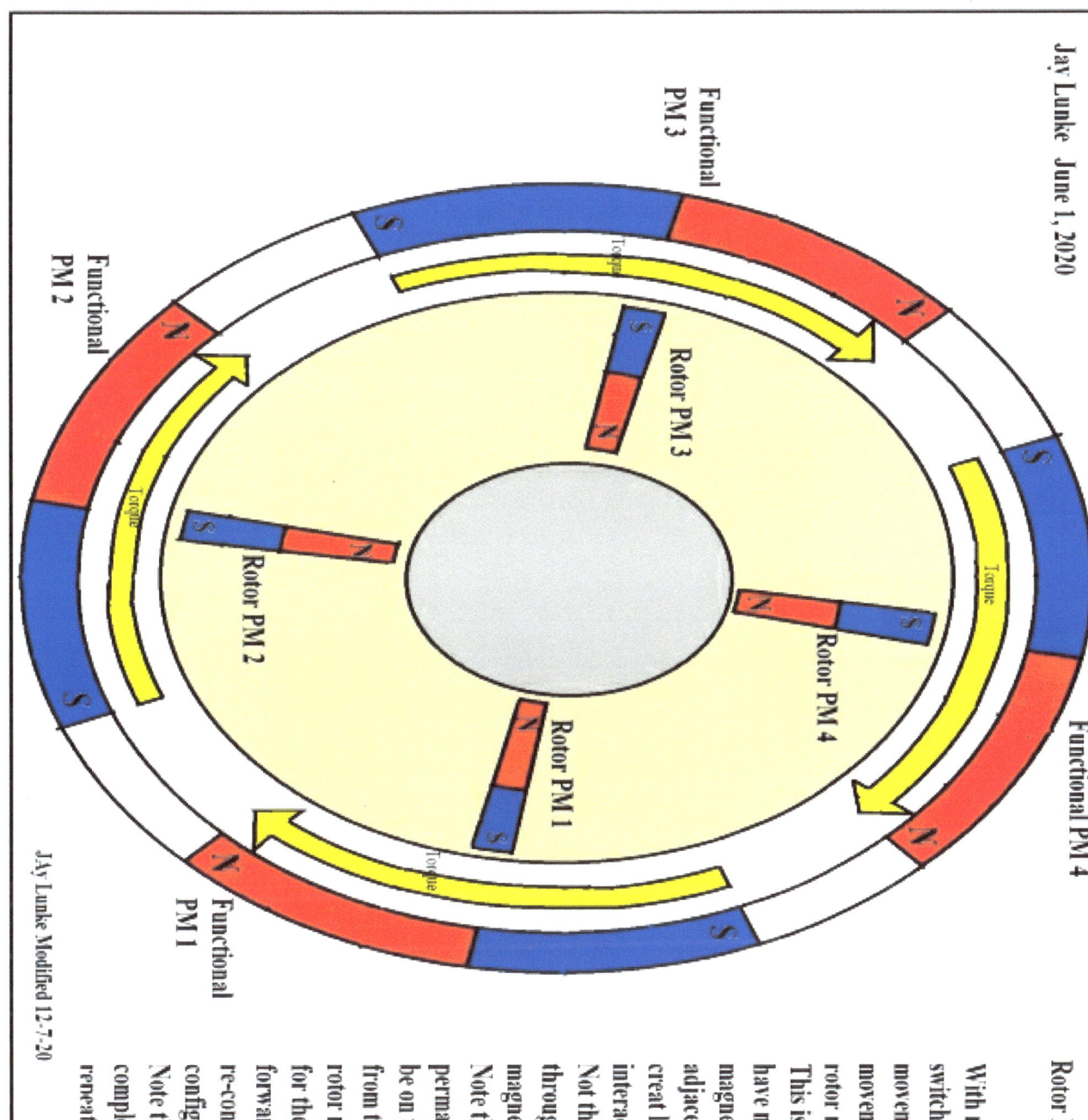

Rotor Movement; Position 4

Jay Lunke Modified 12-7-20

With rotor movement position four, the four switching position that were inactive in rotor movement two have become active in rotor movement four. The four that were active in rotor movement two have now become inactive. This is because the rotor permanent magnets have moved to a new position. The active switch magnet positions now become active with the adjacent stator permanent magnets in order to creat larger functional magnets. The rotor now interacts with the functrional stator magnets. Not that in this movement, that the rotor moves through the middle third of this functional magnet.

Note that the shorter in length the stator permanent magnet is, the greater the torque will be on the rotr magnet because more power comes from the poles of the magnets. So when using rotor movement configurations, They, are used for the shortest amount of time in order to keep forwards torque on the rotor befor the stator is re-configured back into movement one or three's configuration again.

Note that after rotor movement four has completed, then then the four rotor movement is repeated again and again.

57

So, the design we just looked at has the stator at the outside of the rotor magnets whereas the disk motor designs have the rotor and stator magnets side by side of each other. Now both of these designs could use different rotor to stator magnet orientations in them and work with the different configurations. Now each design has their advantages and disadvantages for different applications.

Now another design option has more of a 45-degree angle to it through the rotor. I have to admit that I saw another design that operated with an angle at close to a 45-degree angle in order to reduce the effects of back EMF. So, this is an attempt to increase the torque output with a constant input power. I put this in as another option to be evaluated by people who are smarter than I am.

It may just be that it takes several changes to the current way we design and build motors and generators to get to a design that takes off in the world a coil great free energy device. One a free energy device is proven and built into production and accepted in the world, then many free energy inventions will enter into the world instead of being suppressed.

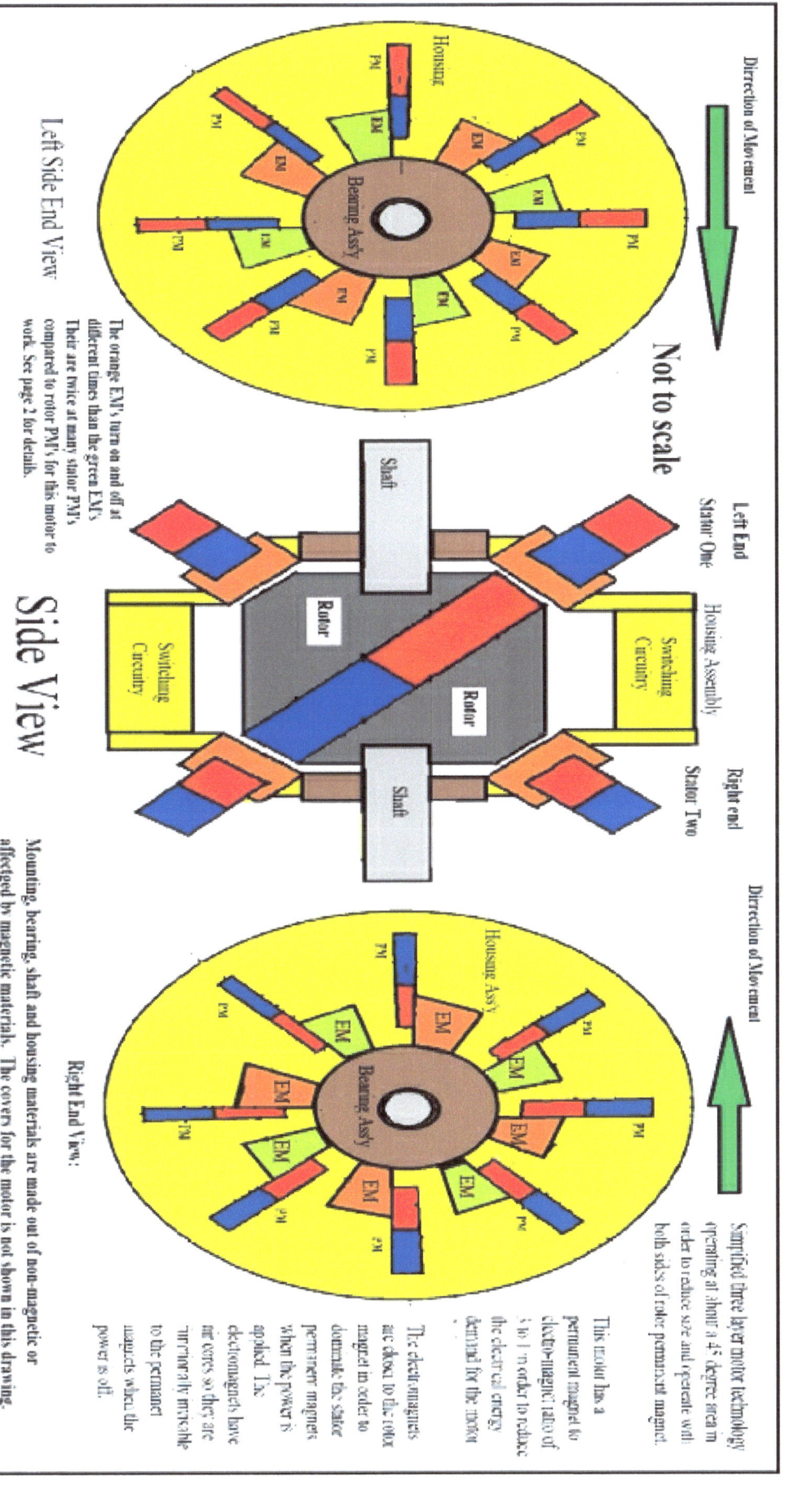

Now the above drawing shows something that is different than the other motor designs I have shown in this book so far. That is the offset of the electromagnets to the permanent magnets in the stator assemblies. What this does is to ensure that when the power is applied to the motor, that the motor will rotate immediately in the desired direction of rotation. After the power is removed from the electromagnets, the two adjacent permanent magnets will have the same torque on the rotor permanent magnets on some of the motor designs. If there is no kinetic energy in the rotor, then the rotation may go in the wrong direction. This should normally not happen because after the electromagnets are powered down, the kinetic energy moves the rotor in the correct position. This design could have applications over the other motor designs. I have it here in order to show you that this new technology has an almost unlimited design option to it.

SUMMARY OF MOTOR APPLICATION

Now since the torque from permanent magnets does not require external energy in order to provide torque in motors and since the ratio of permanent magnets to electromagnets are up to five to one and since some of the energy that generates the flux in the electromagnet is recaptured into the tank circuit and reused again the next time the electromagnet is used in the circuit; the result is most likely to be an over unity system. Again, the energy is from harvesting the torque from the permanent magnets functioning like a battery in this new technology. Now if this is true, then it makes sense to have a highly efficient generator in the system so that the output of the generator will be enough power to power the motor and power circuit at the same time as have torque from the rotor to produce mechanical work. There may even be power left over from the generator to power other devices. Again, these are all theories that have not been proven out yet. Just think if you were the person to build and prove the technology to the rest of the world. How would that make you feel?

POWER GENERATION

There are many generator options that can be used with this new technology that are old and new concepts as well. Pick your own favorite power generation to use.

The first drawing shows how the back side of the rotors can be used in a generator circuit. So, the permanent magnets in the rotors would have dual functions in order to drive the motor at the same time as drive the flux into the generator coils.

The motor uses the Three Layer Electromechanical Movement Technology In the motor section of this system.

The generator is a conventional design in this system. For a system to be a free energy system. You only need either the motor or the generator to be a free energy device.

A free energy system that includes the motor and generator has many potential applications.

Next you will want a system to include a generator that produces 50Hz or 60Hz so that you can use all of your current electrical devices. There are many off the shelf devices that can be used in those systems.

PERMANENT MAGNET TORQUE HARVESTING

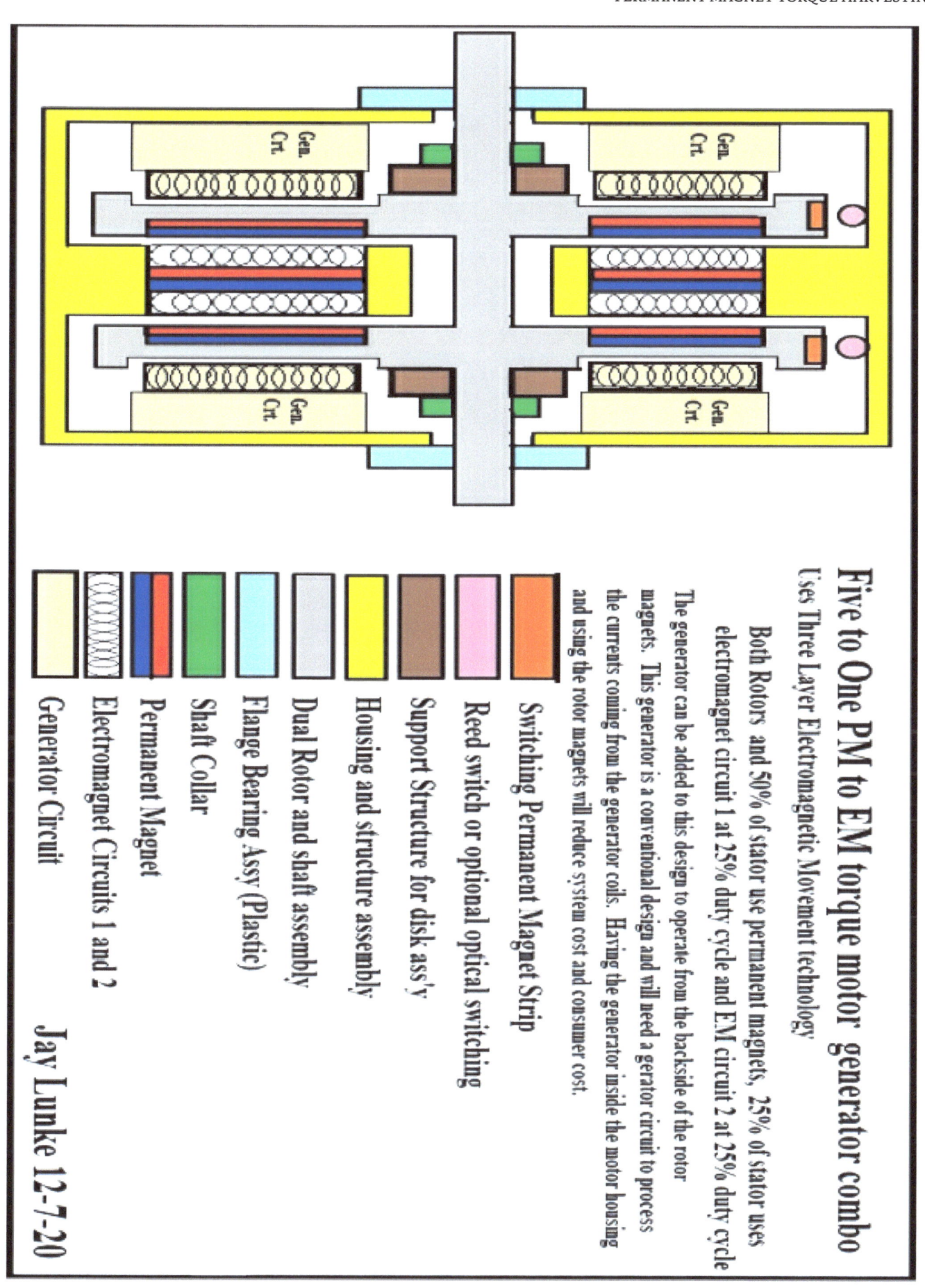

Five to One PM to EM torque motor generator combo
Uses Three Layer Electromagnetic Movement technology

Both Rotors and 50% of stator use permanent magnets, 25% of stator uses electromagnet circuit 1 at 25% duty cycle and EM circuit 2 at 25% duty cycle

The generator can be added to this design to operate from the backside of the rotor magnets. This generator is a conventional design and will need a gerator circuit to process the currents coming from the generator coils. Having the generator inside the motor housing and using the rotor magnets will reduce system cost and consumer cost.

- Switching Permanent Magnet Strip
- Reed switch or optional optical switching
- Support Structure for disk ass'y
- Housing and structure assembly
- Dual Rotor and shaft assembly
- Flange Bearing Assy (Plastic)
- Shaft Collar
- Permanent Magnet
- Electromagnet Circuits 1 and 2
- Generator Circuit

Jay Lunke 12-7-20

The following drawing is a true DC generator. The pick up coil moves 360 degrees in 360 degrees of a magnetic field. This should reduce the back EMF since the magnetic field is built inside of the coil that moves around in a circle. The faster the coil rotates inside the permanent magnets, the larger the voltage will be. This design will need to be built and tested against other generator designs for it's efficiency. I do see a true DC generator as having some advantages over other generators for powering DC circuits or powering DC switched power circuits.

Any generator I show in this book needs rotation in order to produce electrical energy. So there would be a need for an external power source in order to start the motor and generator up to the operating speed. This could be done by having a battery in the system.

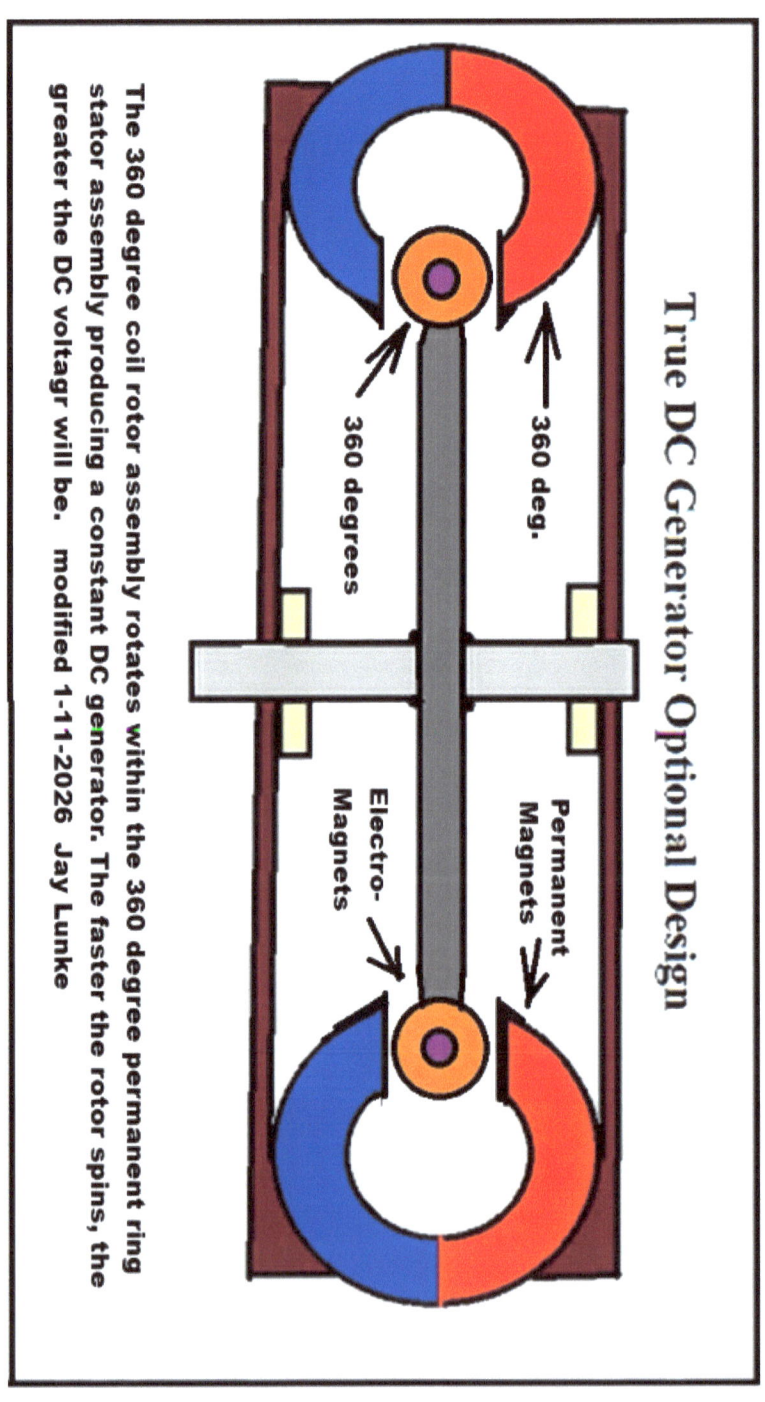

The following concept is an indirect approach in electrical energy generation. The drawing shows a three-legged transformer in the drawing. Now there have been other people working with three-legged transformer designs. Most of those designs are stationary. Dr. Smith has configurations that are similar to my design. So, you would need to get legal advice before using this design in any products that you would intend to build and sell.

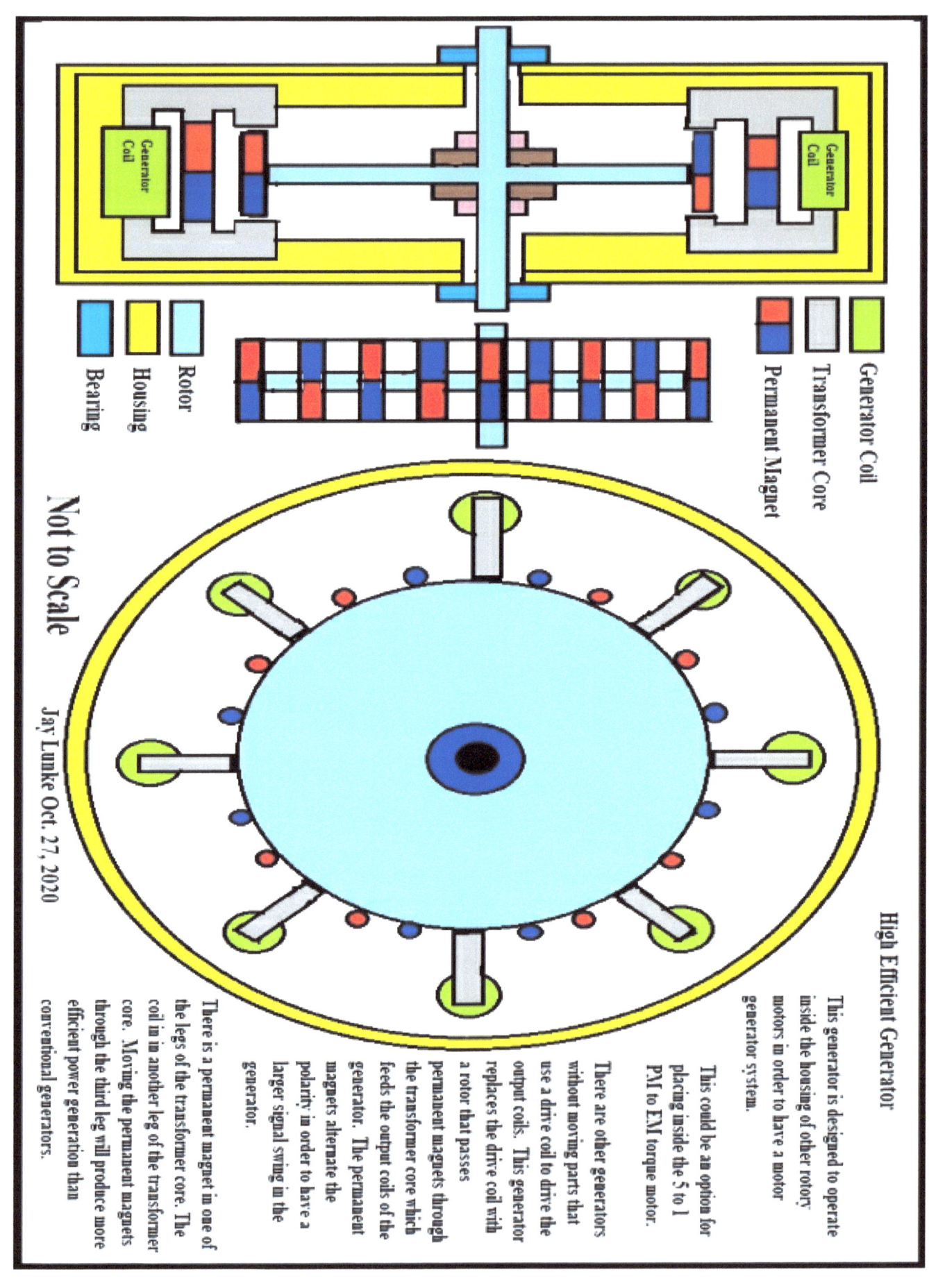

A two-legged transformer can be used as well where one leg would be the output coil and the other leg would have the permanent magnets passing through it. Each following permanent magnet would have the poles in the opposite direction from the magnet that proceeded it. It would be best to test both configurations and use the one that performs the best. But as with the motors, the generators need to be evaluated to make sure that there are no patent infringements in using them. It is my opinion, that if you found out that someone had the rights for the design, that they would be grateful for the opportunity to share a percentage of a royalty with you. Again, this is just my opinion, you must seek legal advice in that area.

There is also the possibility that the patent has expired. You could also open source the technology so everyone has the opportunity to use the technology.

My designs are a starting point for you to improve my designs. Most inventions started by people learning from other people before them. The is a joy in having people seeing your work, and them telling you that from that work they came up with an even better design or technology.

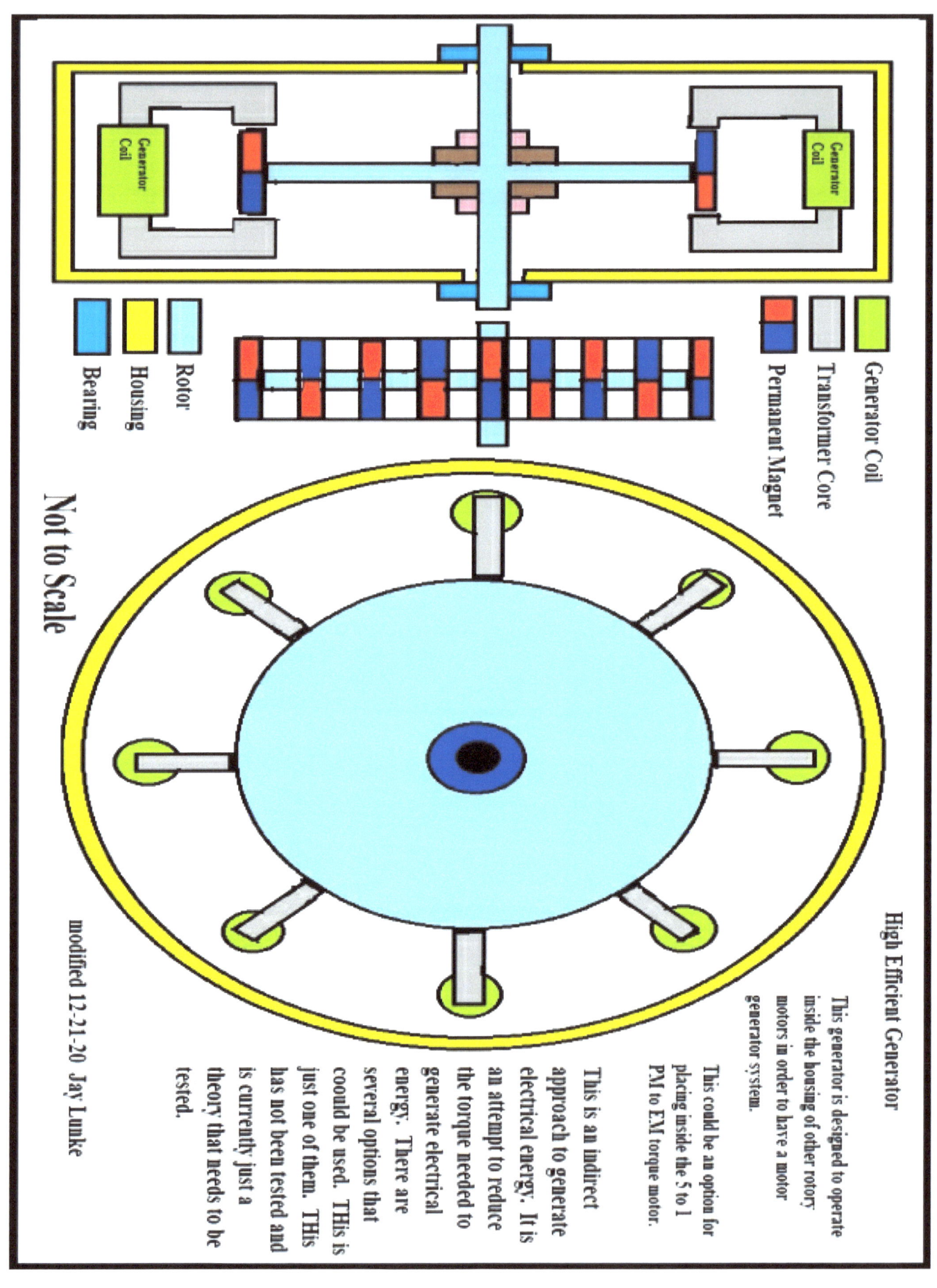

Now the following drawing is a unique thought I had in power production. But I have not performed a search through other generators because I want all of my designs to be public domain. What I like about this design is that if the 5:1 PM to EM motors were built on much larger disk assemblies, then this generator could be incorporated into the same disks as the motor. This would be very easy to incorporate into the design that I do not even have to create a new drawing for it.

The basic goal of a highly efficient generator is to reduce the torque needed to turn the generator under load as close to nothing as possible. Again, there are other generators claiming even better performance than that. One design claim is that when the generator is operating, that the motor will speed up because the generator provides forward torque to the motor while it is operating. If that is true, then by all means ask for permission to use their design.

In the meantime, this is just another option to reduce torque of electrical power generation. Again, I have not tested any of the generator designs. These are all theories of operation.

In the drawing below the permanent magnets are on a different disk. The permanent magnets cross the center half of the electromagnets. The poles of the electromagnets are not close to the poles of the permanent magnets. The hope is that the combination of the angles of the components and the poles of the components would decrease the torque of the generator reducing the generator shaft speed causing the generator to produce more electrical energy than a conventional generator.

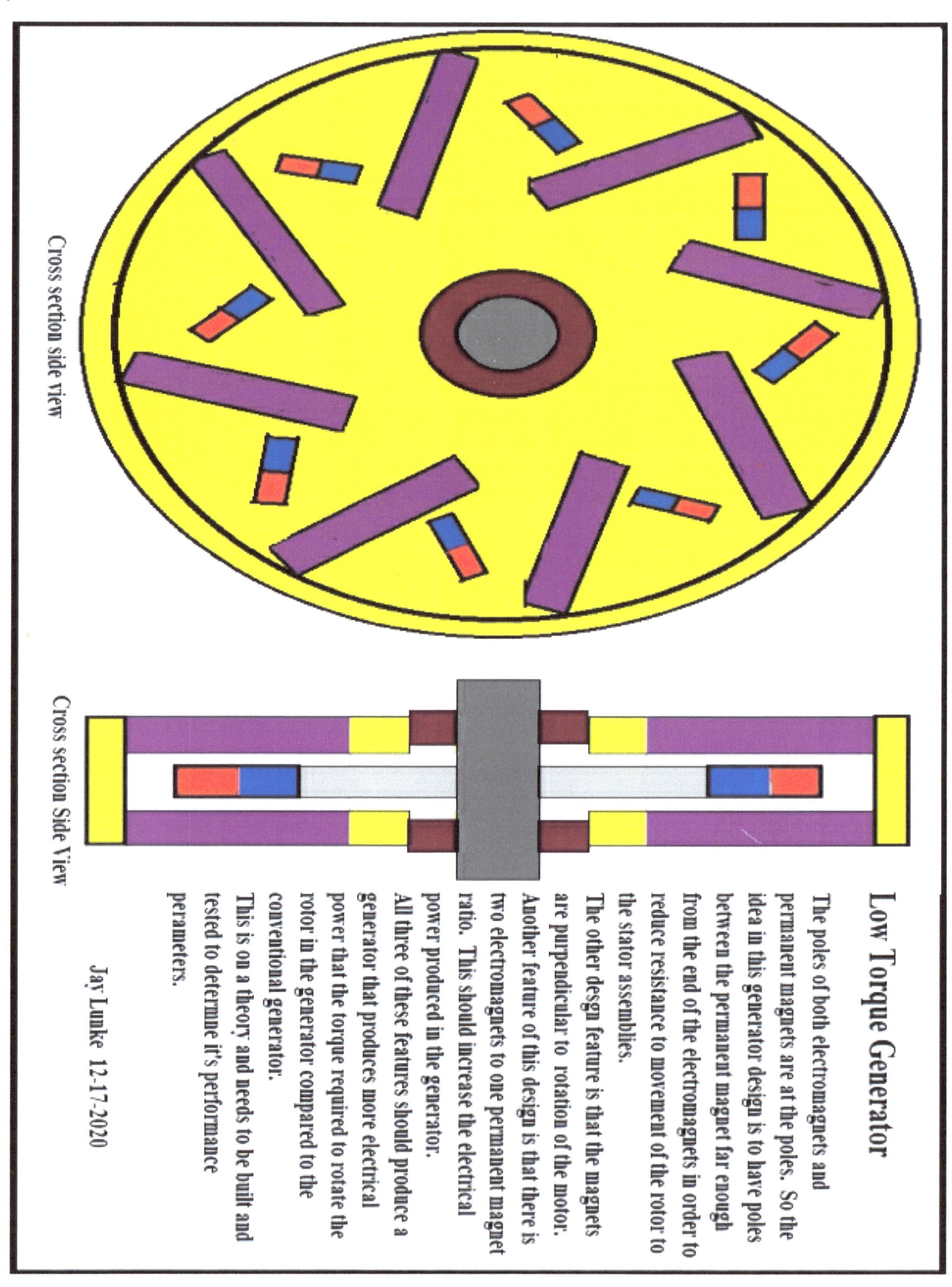

Cross section side view

Cross section Side View

Low Torque Generator

The poles of both electromagnets and permanent magnets are at the poles. So the idea in this generator design is to have poles between the permanent magnet far enough from the end of the electromagnets in order to reduce resistance to movement of the rotor to the stator assemblies.

Another feature of this design is that there is two electromagnets to one permanent magnet ratio. This should increase the electrical power produced in the generator.

The other design feature is that the magnets are purpendicular to rotation of the motor.

All three of these features should produce a generator that produces more electrical power that the torque required to rotate the rotor in the generator compared to the conventional generator.

This is on a theory and needs to be built and tested to determine it's performance perameters.

Jay Lunke 12-17-2020

THE BIG QUESTION IS WHY USE AN INDIRECT METHOD TO GENERATE ELECTRICAL POWER?

The reason is that the interactions between the two systems are different in nature. I will not speak a lot about the conventional generation methods. The coils in conventional electrical generation create a magnetic field that opposes the permanent magnet that is passing the electromagnet. This opposition reduces the efficiency of the power generation. In the indirect transformer generator design, the moving permanent magnet passes between two metal posts creating another leg of the transformer. Now as a permanent magnet is passed by a metal post, the permanent magnet will have attraction to that post, that attraction creates forward torque on the rotor and on the other side of the post it creates negative torque for the rotation of the rotor. So, when a vector analysis is performed on these torques, a large part of the torques cancel themselves out. Not all of it though. So, it would be good for someone to build one and compare the performance of it to other generators. So, in the three-legged transformer design, one permanent magnet is already in the circuit. It has flux going through the generator coil. Now electrical generation occurs only when there is a changing flux through that coil. So, when the permanent magnet passes through the third leg of the transformer, most of the flux in the coil changes in strength producing electrical energy. Depending on the polarity of the rotor permanent magnet, flux will either be added to or taken from the leg of the transformer that has the generator coil in it. Now if the three-legged transformer configuration has no advantage to conventional electrical energy generation, then a two-legged transformer configuration can be used. The two-legged configuration would have the generator coil on one leg and the passing rotor permanent magnet on the other leg. If the two-legged transformer configuration has the same performance as the three-legged transformer, then use the two-legged transformer configuration to save money in the final device.

If the motor and tank circuit system is self-sufficient to operate, then using designs that have the motor and generator in them is the way to go to reduce the final cost of the system.

Now once the best on board or in this case on device generator configuration has been developed, then you want to incorporate that design into a motor package that can be mounted into the largest variety of design applications. The following drawing has a motor/generator combination with several mounting surfaces on it in order to provide that flexibility.

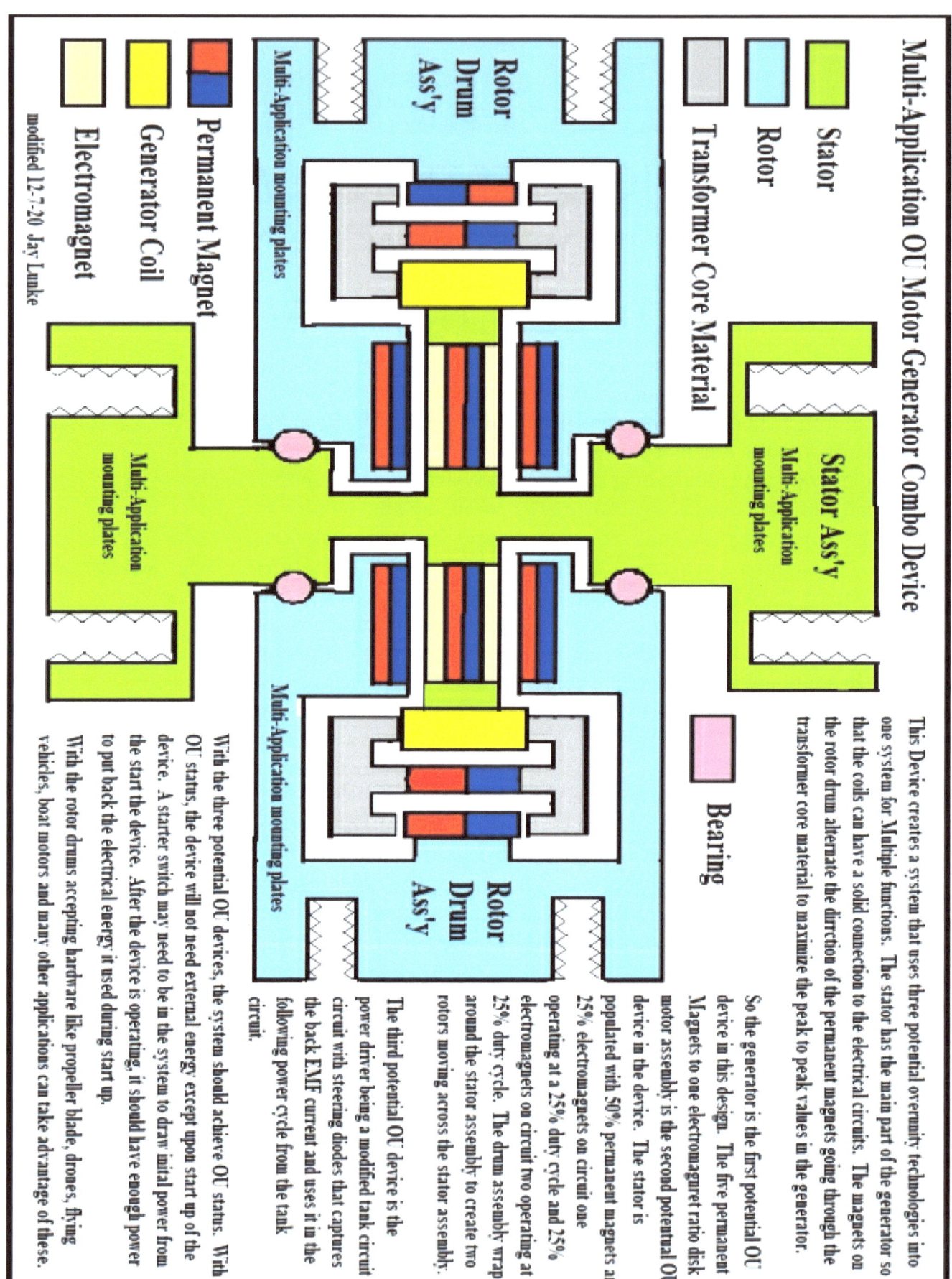

The drawing above has the three legged transformer design in it while the following drawing shows a two legged transformer generator design in it.

I show both drawings because I do not know which design will work better. When designing a a new technology, it is best to have several options going into the build. For one, you do not want to build one design and then find out that the other design was a better design and then someone else patents the other design before you did. I hurd that in the USA, the first to patent is the new way patents are giving out today. Some people do not like patents because minor improvements to a design can be a reason for someone else to get a patent on an invention. Of course if it was me, I would use the open source route.

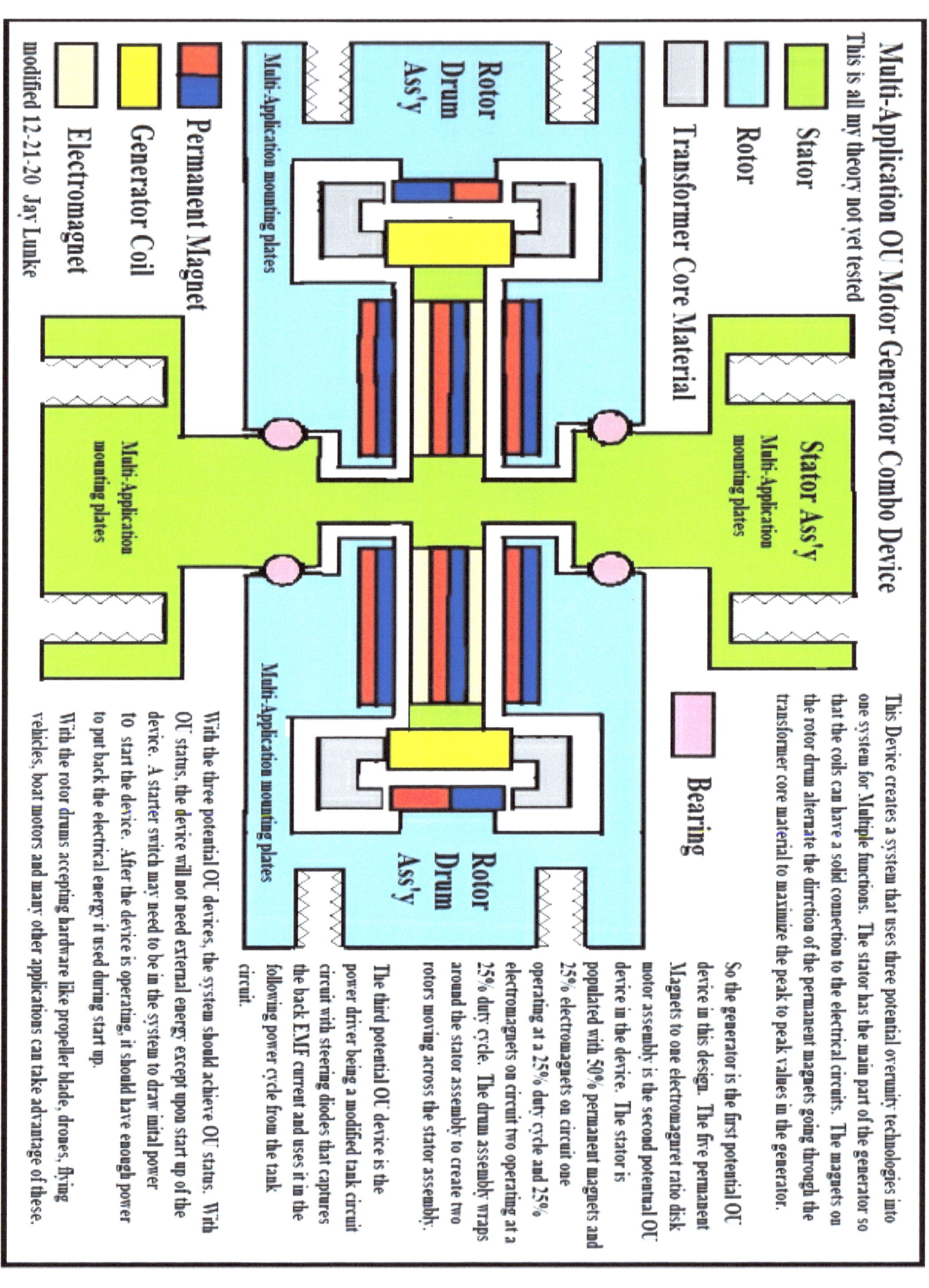

Notice that the transformer has the coil on the stator assembly. The motor designs should be designed without using commutators in them because they are a mechanical assembly that wears out. This new technology needs to be designed for as low maintenance as possible.

Also notice that the drawing above has larger motor magnets in it. The motor space should be larger than the generator space, when possible, in order to have more power from the motor.

Another thing that you need to know is that the farther out on the circumference the motor is, the more torque the motor will have from it. So, you may want to design your motor with the generator on the inside of the disk and the motor on the outside of the disk in order to take advantage of the higher torque from the design.

As in the generator drawing, if the three-legged transformer does not demonstrate better performance than the two-legged transformer configuration, then the two-legged configuration should be used in the design. Now there are unlimited applications for high efficiency magnetic motors. There are so many advantages over the gas engine. I personally have an all-electric house and an electric car because I want to be active in having a greener world around me. The following is one example of using the new motor designs.

The flying car is not a new idea, so the design you have seen, if not in flying cars, then in drones. These will become more common place once high torque over unity, or should I say magnet torque powered, motors are available. It is amazing how many cartoons will have a technology that eventually becomes a reality later on in life. The flying cars may not be allowed to fly very high off from the ground, but some day they will be here in volume.

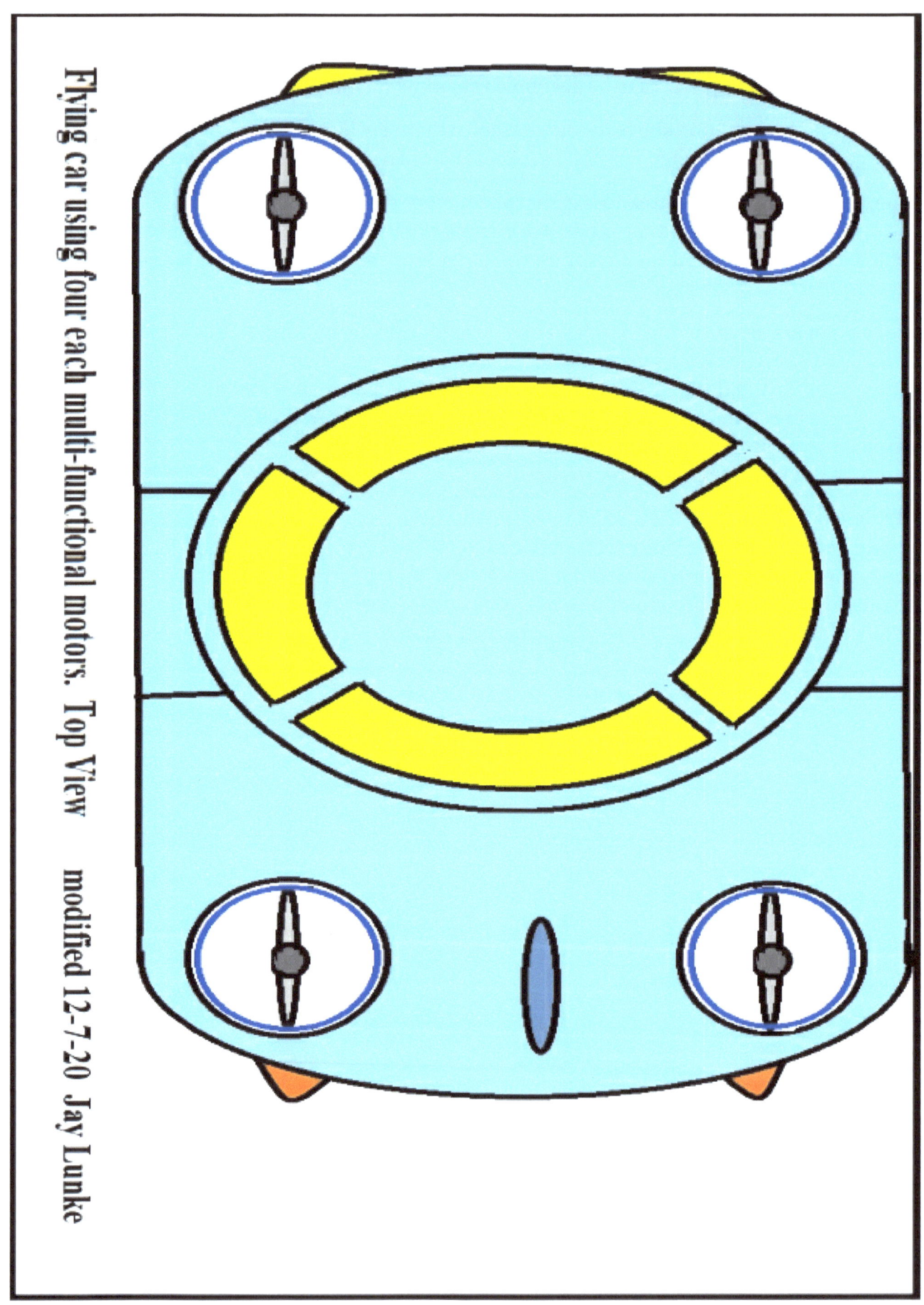

Flying car using four each multi-functional motors. Top View   modified 12-7-20 Jay Lunke

In reality, the motors would be much larger in order to allow more air to flow through them. This drawing only shows the concept of such a machine. Many other flight devices use the four-motor design configuration.

The following drawing brings us back to earth or even under the earth. The motor design can pipe crude oil where it is needed to be. It can also move water where it is needed as well.

This can be very helpful for farming where water is currently limited. Large ships and submarines can use this type of motor for propulsion.

What is nice about this design, is that it gets away from using the conventional shaft. This means the motor design can be incorporated into the final application. A washing machine or dryer are two appliances that could have the motor as part of the drum assembly. The more you can incorporate the motor into the final product assembly, you are going to save money in building the product. It also means the product build will save the customer money as well.

High Efficiency Flow Motor
Incorporating 5 to 1 PM to EM Torque Ratio

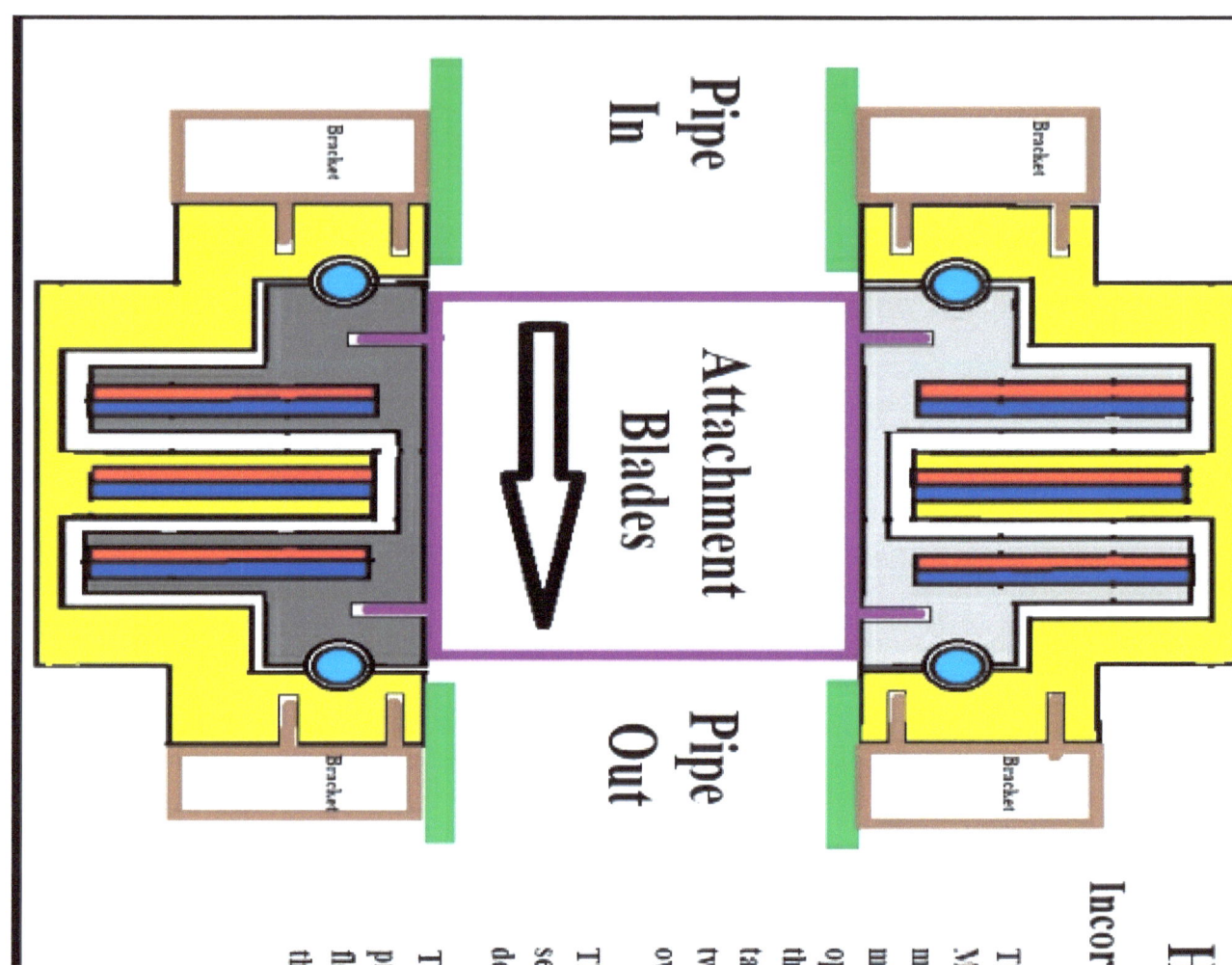

This motor design shows that the Three Layer Eletromechanical Movement Technology has far reaching design potential. This motor has the four segment movements the same as the other motors using this technology. This means the two electromagnets operate at a 25% duty cycle and the motor has forward torque throughout the full 360 degrees of rotation. Also the modified tank circuit with steering diodes will operate this motor. These two design features should be so efficient that it should reach overunity levels.

This motor design has several mounting holes to accomidate several different attachments in order to allow this motor design to be used in more applications for it.

This drawing does not show all of the parts in this design for a production build of the gaskets to seal it from the leaks of the flowing liquid. THe intent of the drawing is to demonstraight the flexability in the designs possible with the technology.

Jay Lunke Oct. 17, 2020

This motor has the ability to place different attachments on the inside in order to accommodate several different applications. The mounting brackets are also adaptable for multiple applications. The inside circumference can become a lot larger for different applications.

The next drawing will show how several of these motors put together in a system can produce a fast either underwater vehicle or they could be placed on a plane like a jet engine.

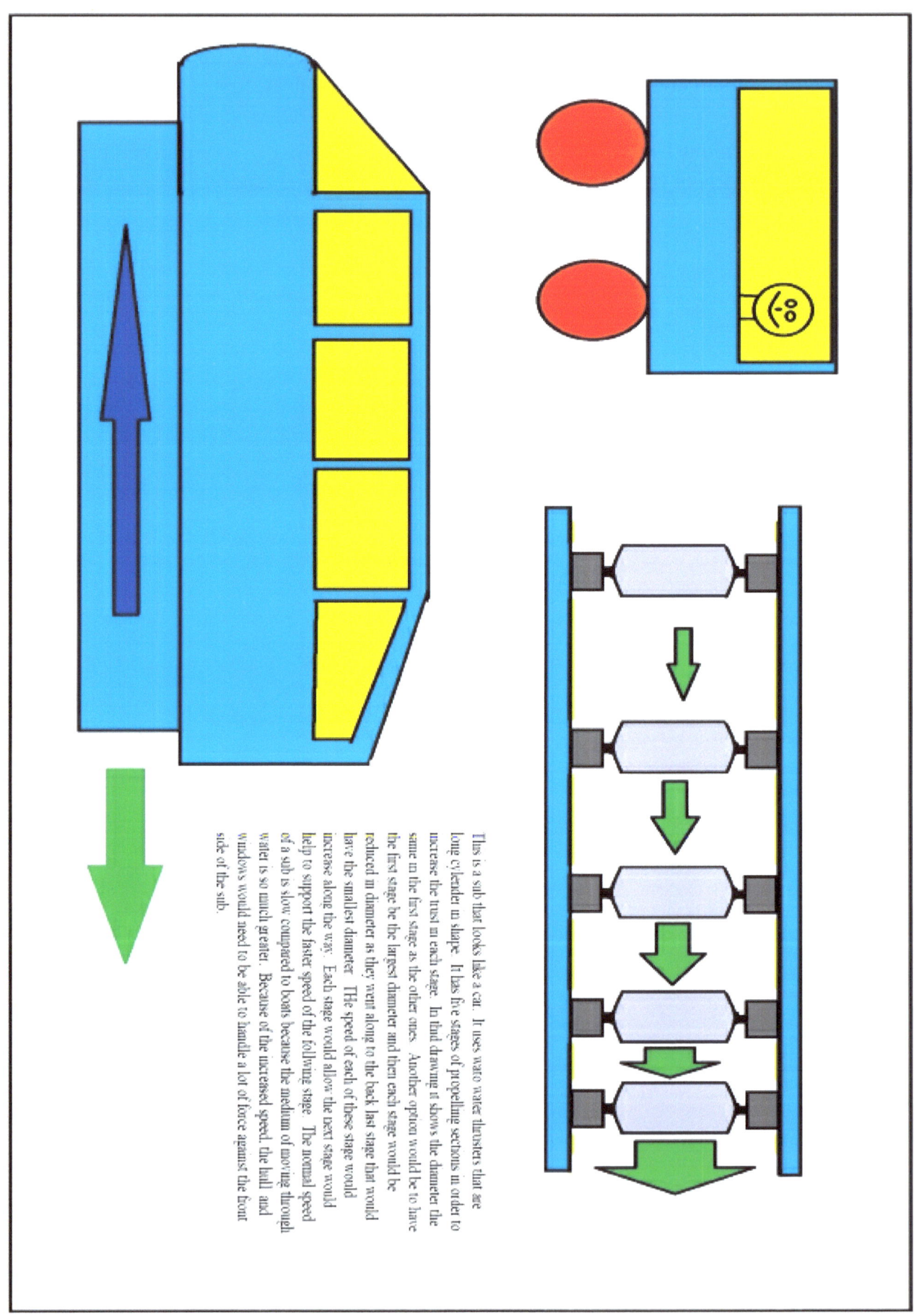

This is a sub that looks like a car. It uses water water thrusters that are long cylender in shape. It has five stages of propelling sections in order to increase the trust in each stage. In titid drawing it shows the diameter the same in the first stage as the other ones. Another option would be to have the first stage be the largest diameter and then each stage would be reduced in diameter as they went along to the back last stage that would have the smallest diameter. The speed of each of these stage would increase along the way. Each stage would allow the next stage would help to support the faster speed of the follwing stage. The normal speed of a sub is slow compared to boats because the medium of moving through water is so much greater. Because of the increased speed, the hull and windows would need to be able to handle a lot of force against the front side of the sub.

The inside circumference can become many times larger for different applications like that of the next drawing.

I got carried away by increasing the inside diameter of the motor when I designed a flying craft that looks like a flying saucer. Now if the flying saucer has not achieved over unity status, then it will need to have the batteries recharged. The next drawing shows you how the wind can be used to recharge the flying craft.

The one nice thing about using electric motors for flying crafts is that with the additions of some seals and water/corrosion protection designed into the motors, they can move from the air into the water and back into the air again. Just think of how nice that would be on vacation.

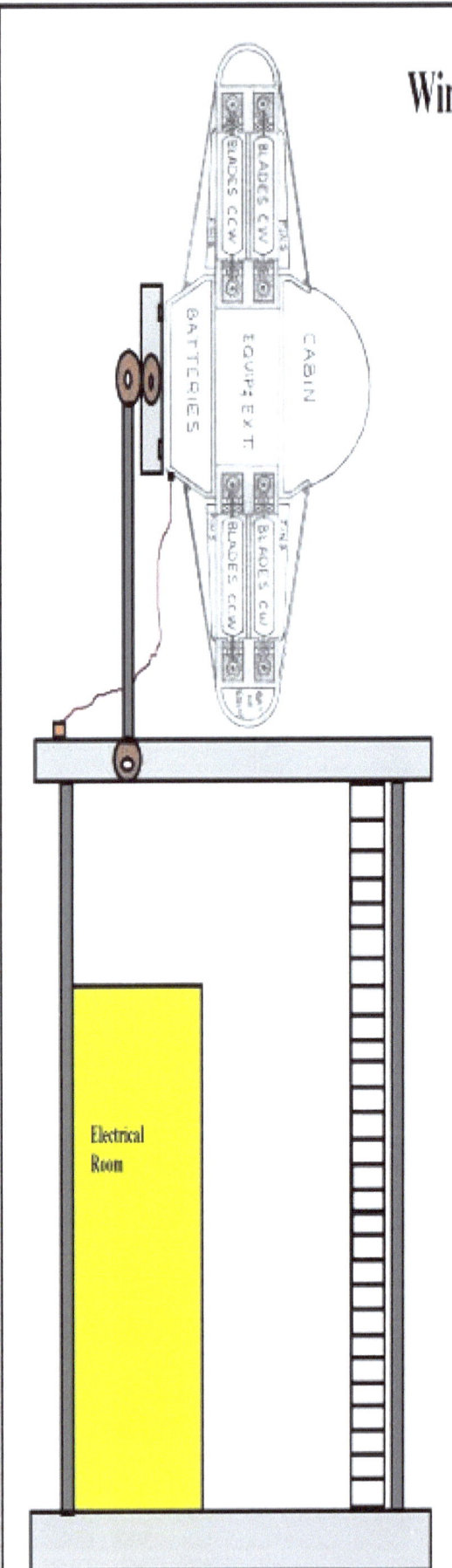

Wind Generator Skycraft Combo

This Skycraft was designed with the flowthrough motor design I came up with in the 1980's. The motor in the skycraft can also be used as a generator. So when you have a pad fore the skycraft raised high enough off from ground level, you can rotate it 90 degrees and use it as a wind generator while it is being parked. The electrical room can store the electrical energy that is generated by the skycraft.

The skycraft has motor/fan assemblies that when in flight act like gyro's. Changing the speed of one gyro more than the other creates the turning movement of the craft. There are fins that dirrect the airflow that generate the forward and reverse movement. Optional side electric motors can be installed to increase the forward movement. Gaskets along with other equipment will allow this skycraft to move around under the water as well. If and when the overunity electric motors are developed, then the battery pack will not be needed on the craft. The craft is simular to a helicopter in that it is limited to operating in the atmosphere. The blades are protested from being damaged from tree branches and other objects.

Many of my designs have the motor and part of the final assembly in order to save the cost of the final product. As you can see, the motor is a part of the skycraft structure. As in electric cars, the range of this craft is limited by the size of the battery. Because this craft does not require a lrge landing strip, it can be operated in many more places than the conventional aircraft. Also being able to move from the sky to the water will bring greater adventures for the people riding in this craft.

It is posible to have several of these connected together like a train in order to move products from one location to another. This can be a good platform for moving into a new world of opertunities.

This size of the skycraft can be greatly increased. Just think if it was large enough to be your home so that when you go to work or on vacation, you bring your home with you.

If you want to know more information on the details of the technology used in the skycraft. I have that information in the book I wrote "A free gift that may be over unity or free energy to the world"

There is a free download of version 1 on the web. I also give a free kindle version 5 times every three months. I also have it at a very low price. I want to help to create a better world for all of use. I am not looking for money but to share what I know with other people.

The Lunkster

Now these motor designs do have their limitations. The motors can not become to hot or the magnets will loose their magnetism. The motors will not provide thrust in a vacuum like that of space. This technology is designed for larger motors. So hand tools will have their current motors in them.

The motors will be more expensive because they use more permanent magnets in them. That is where they get the torque from. The life of the rare earth magnets are several years. Think of how often you have to replace batteries in devices. The rare earth permanent magnets will not have that problem.

BUILDING A PROTOTYPE MOTOR FOR THE HOBBYIST

Now let's get back to this present day. There are things that have to be done first with this technology to prove it out before we can proceed to fly in the sky with it. We need to optimize different aspects of the technology in order to have proper tools and data to build the worlds best motors. One very important thing is the electromagnet design. What size wire gauge to use for the motor. How many times does it need to be wound? The size and shape of the coil. The proper way to attach it to the plate assembly. The distance of the coil to the rotor permanent magnets along with the distance of the rotor permanent magnets to the stator permanent magnets. I have seen people performing copycat designs that are functional, but not the most efficient. So, performing testing all along the way of the design is important. There is also the conflict of the design engineer wanting to build the device to the best it can ever be with the manager wanting the largest profit as can be. With this new technology, I hope the design engineer wins that battle.

So, the following prototype design is mainly to get the correct ratio of the stator electromagnets and stator permanent magnets in the correct placement to the rotor permanent magnets. This is because this new technology does not have this data yet. Many of the other components have been well tested.

PERMANENT MAGNET TORQUE HARVESTING

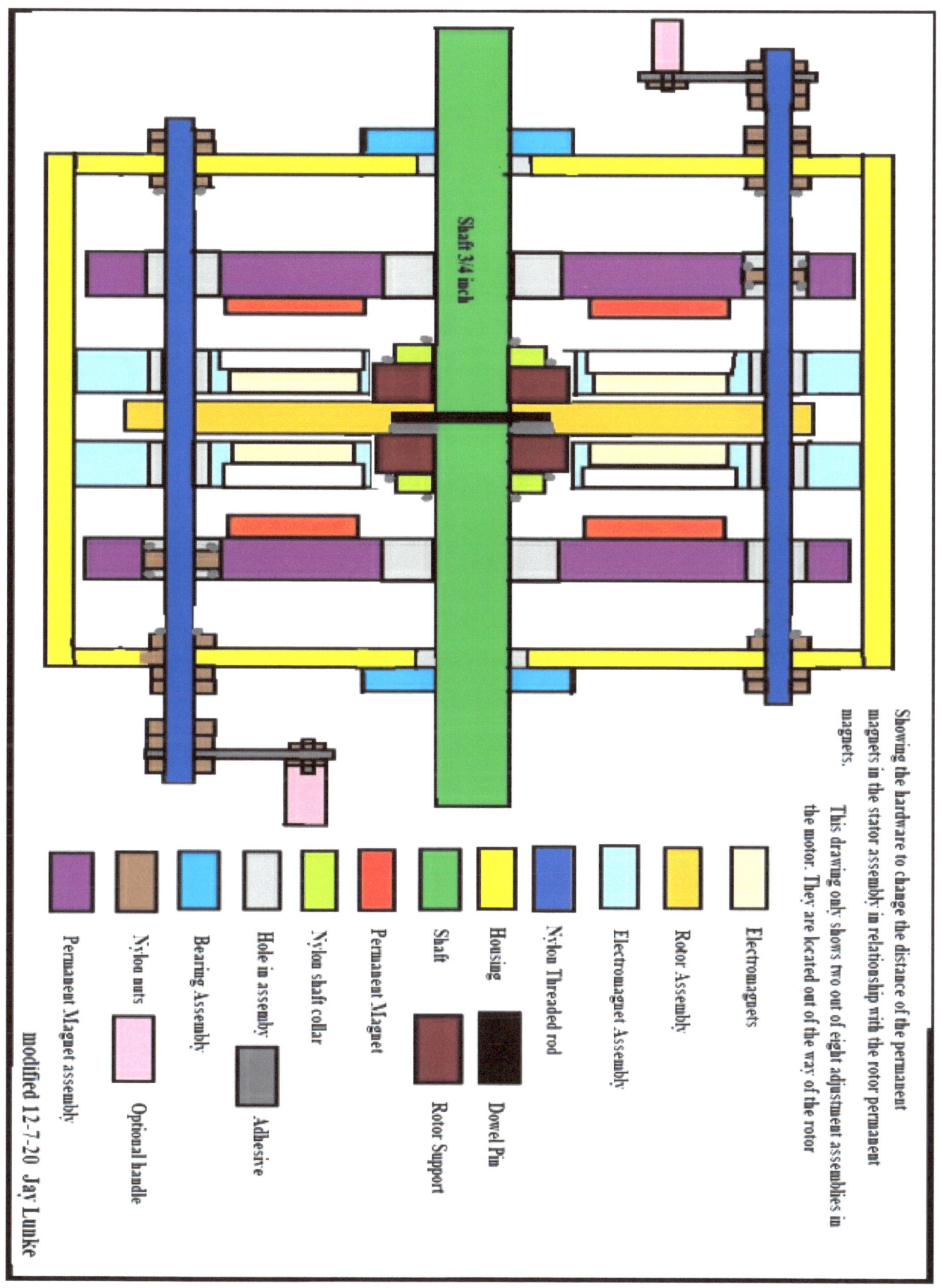

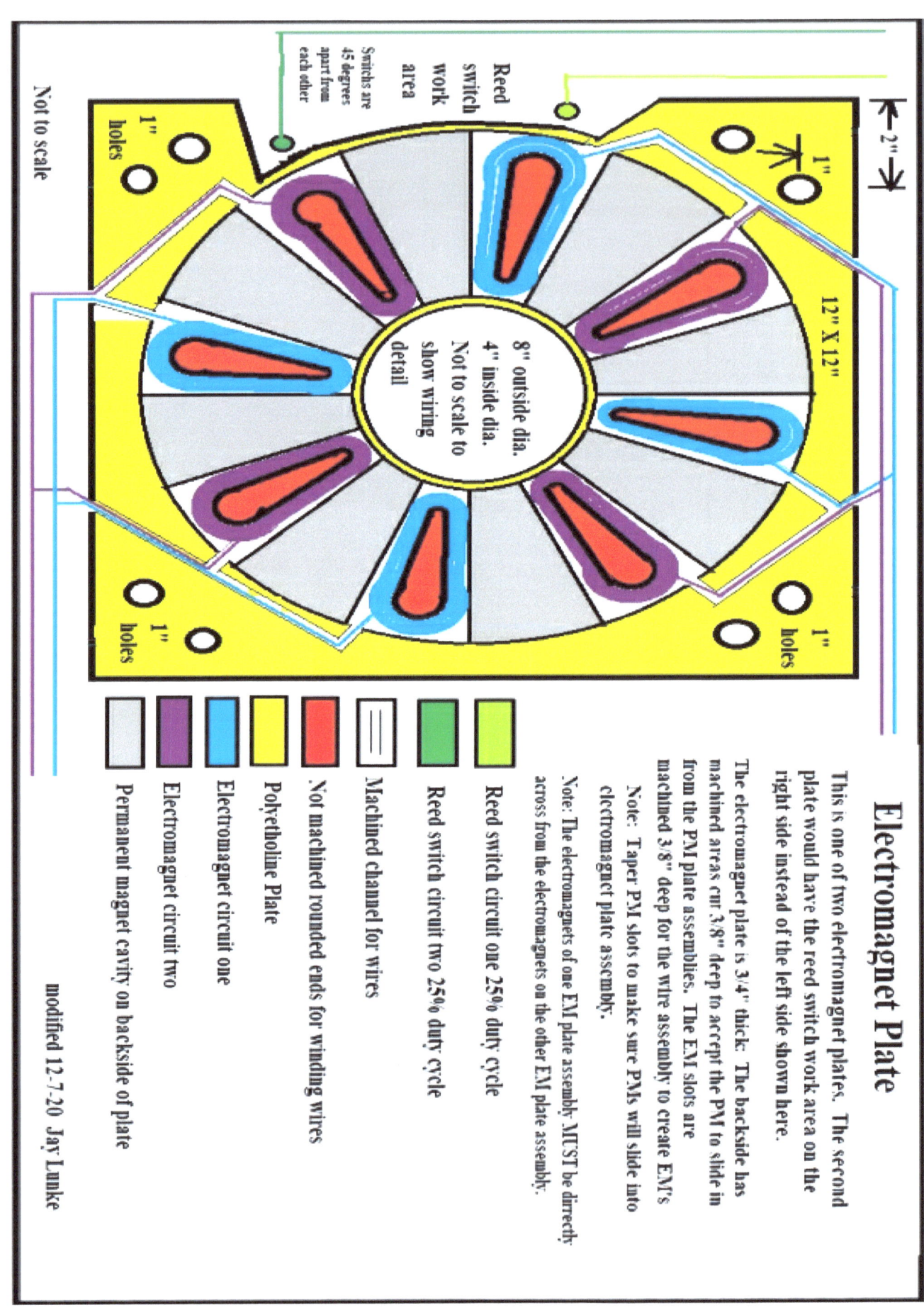

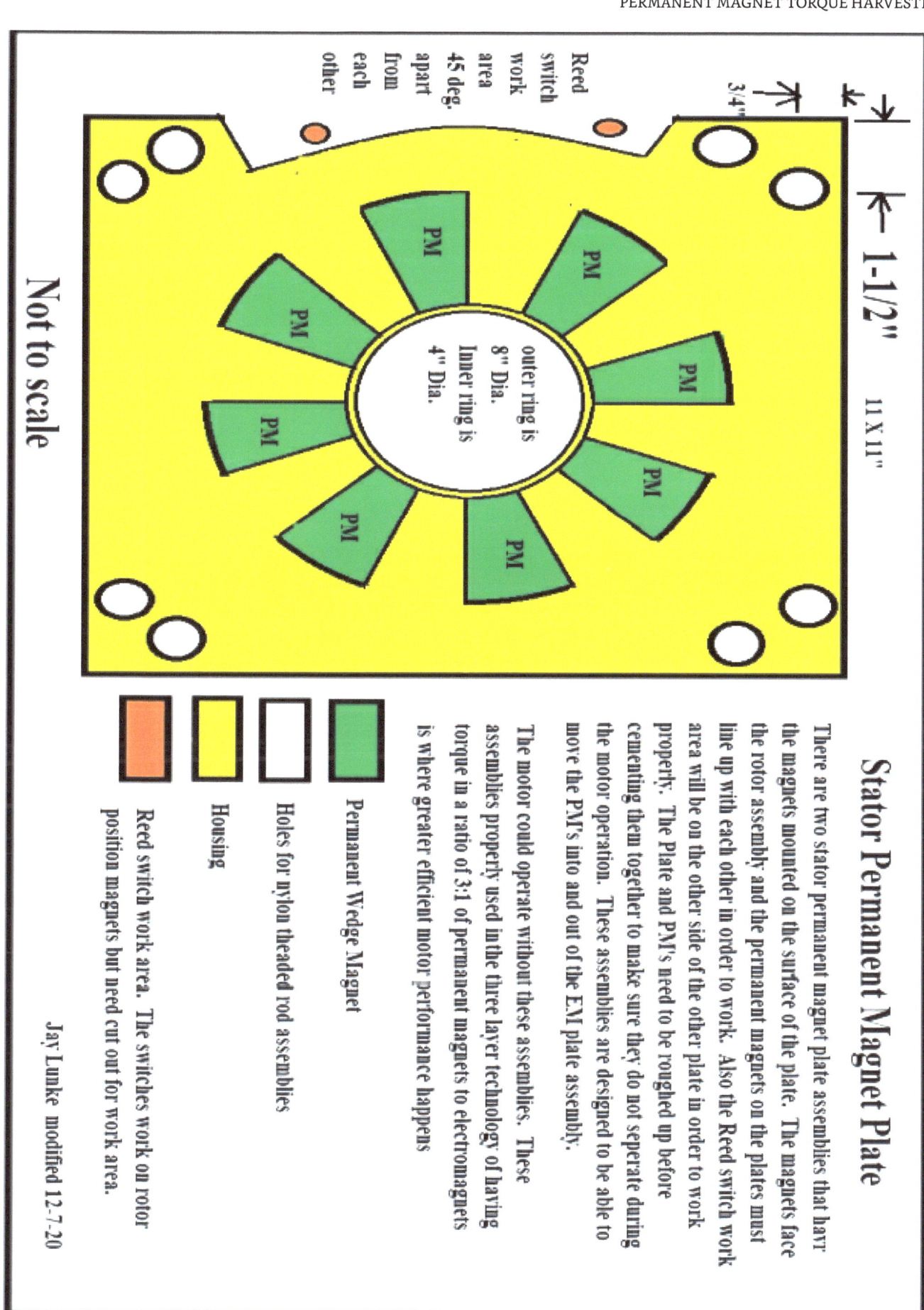

Stator Permanent Magnet Plate

There are two stator permanent magnet plate assemblies that have the magnets mounted on the surface of the plate. The magnets face the rotor assembly and the permanent magnets on the plates must line up with each other in order to work. Also the Reed switch work area will be on the other side of the other plate in order to work properly. The Plate and PM's need to be roughed up before cementing them together to make sure they do not seperate during the motor operation. These assemblies are designed to be able to move the PM's into and out of the EM plate assembly.

The motor could operate without these assemblies. These assemblies properly used in the three layer technology of having torque in a ratio of 3:1 of permanent magnets to electromagnets is where greater efficient motor performance happens

- Permanent Wedge Magnet
- Holes for nylon theaded rod assemblies
- Housing
- Reed switch work area. The switches work on rotor position magnets but need cut out for work area.

Jay Lunke modified 12-7-20

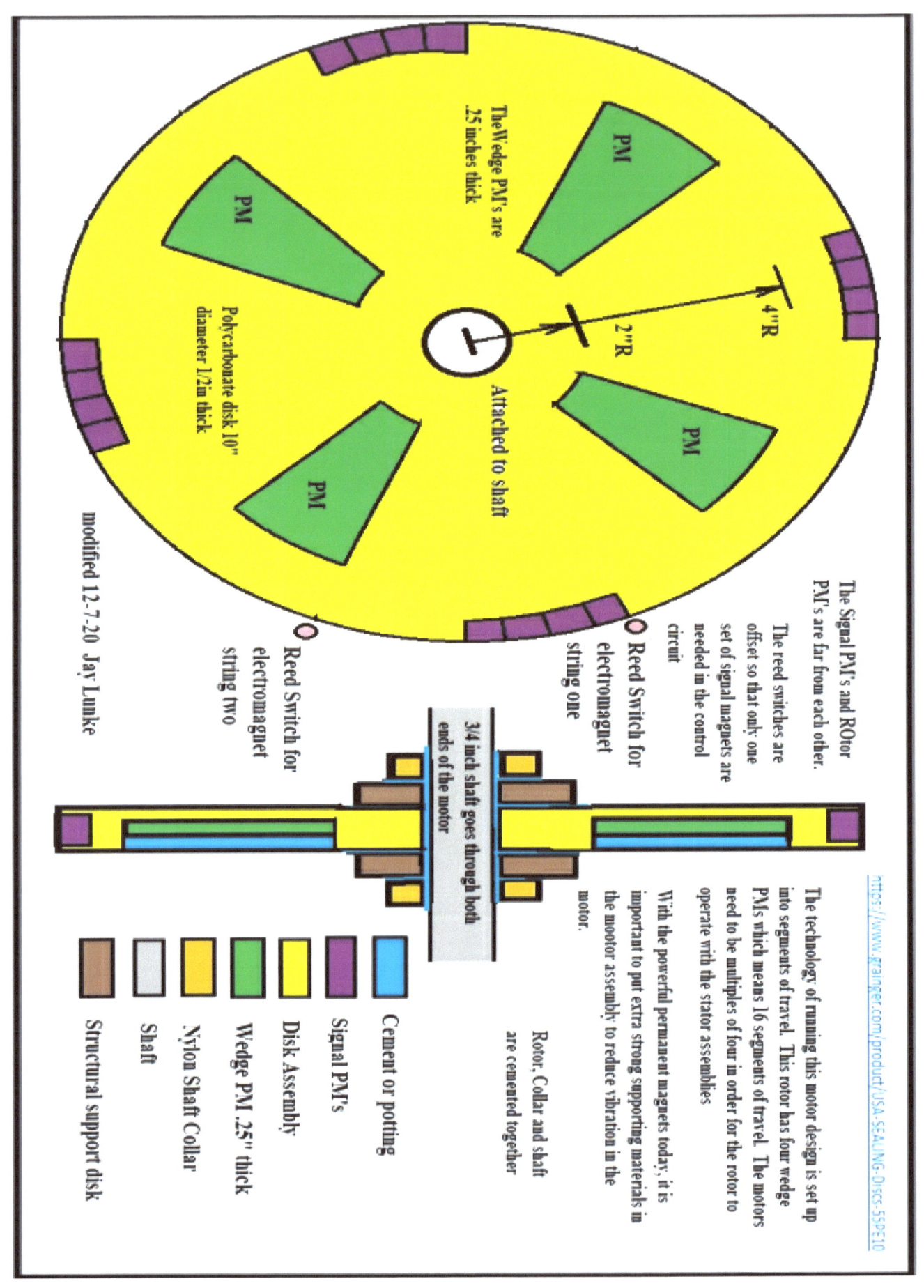

Stator Location Rods

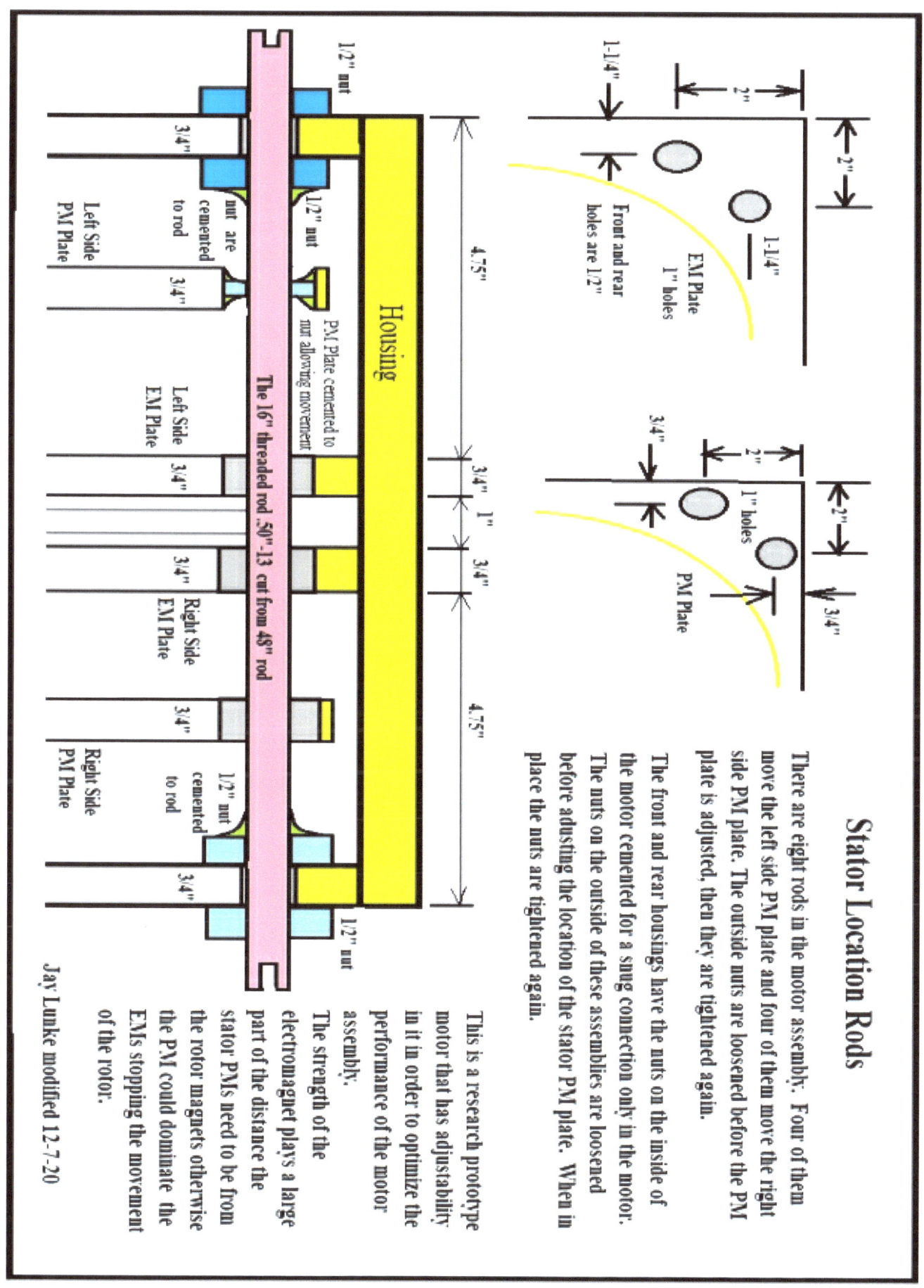

There are eight rods in the motor assembly. Four of them move the left side PM plate and four of them move the right side PM plate. The outside nuts are loosened before the PM plate is adjusted, then they are tightened again.

The front and rear housings have the nuts on the inside of the motor cemented for a snug connection only in the motor. The nuts on the outside of these assemblies are loosened before adjusting the location of the stator PM plate. When in place the nuts are tightened again.

This is a research prototype motor that has adjustability in it in order to optimize the performance of the motor assembly.

The strength of the electromagnet plays a large part of the distance the rotor PMs need to be from the rotor magnets otherwise the PM could dominate the EMs stopping the movement of the rotor.

Jay Lunke modified 12-7-20

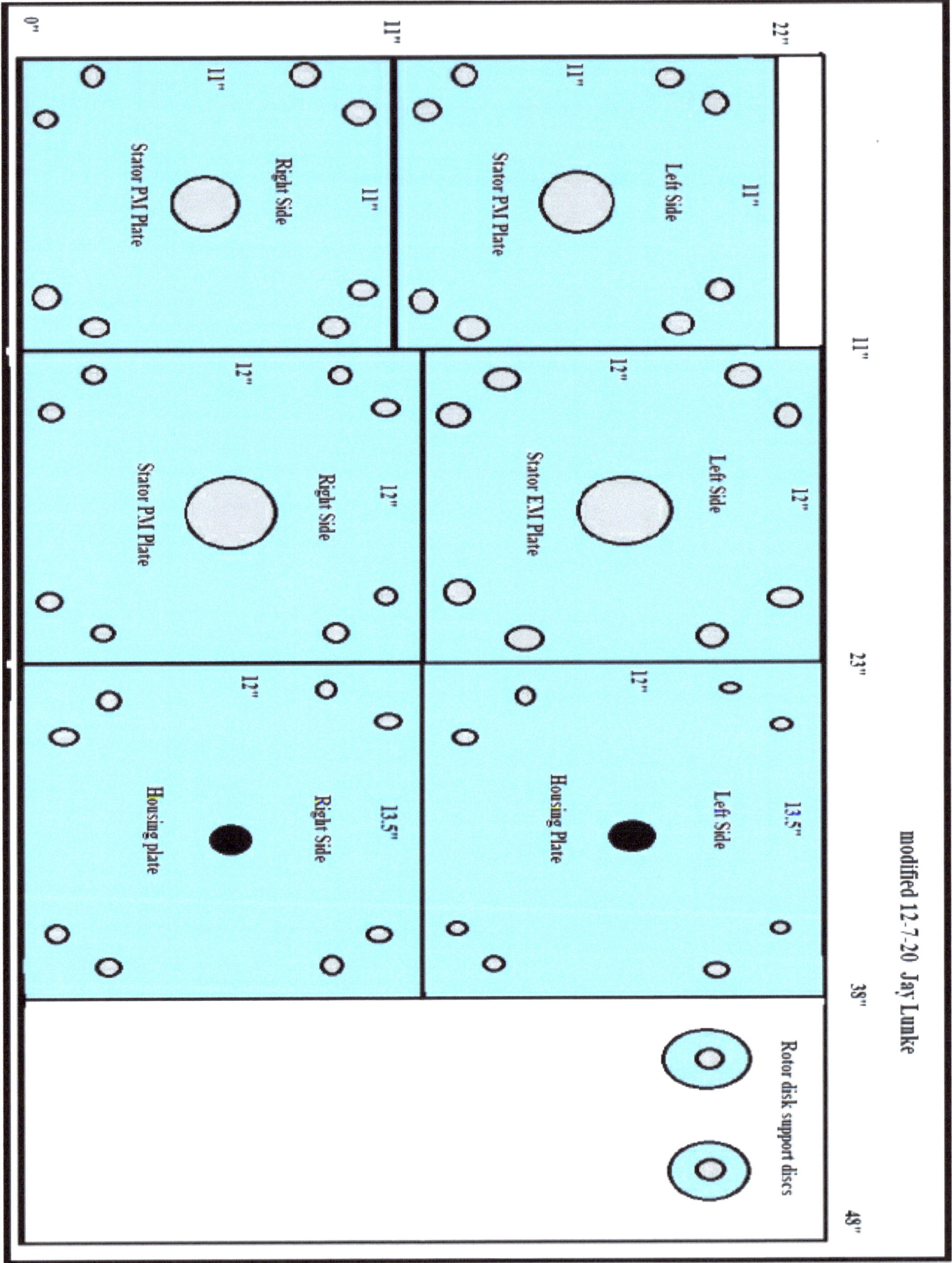

PERMANENT MAGNET TORQUE HARVESTING

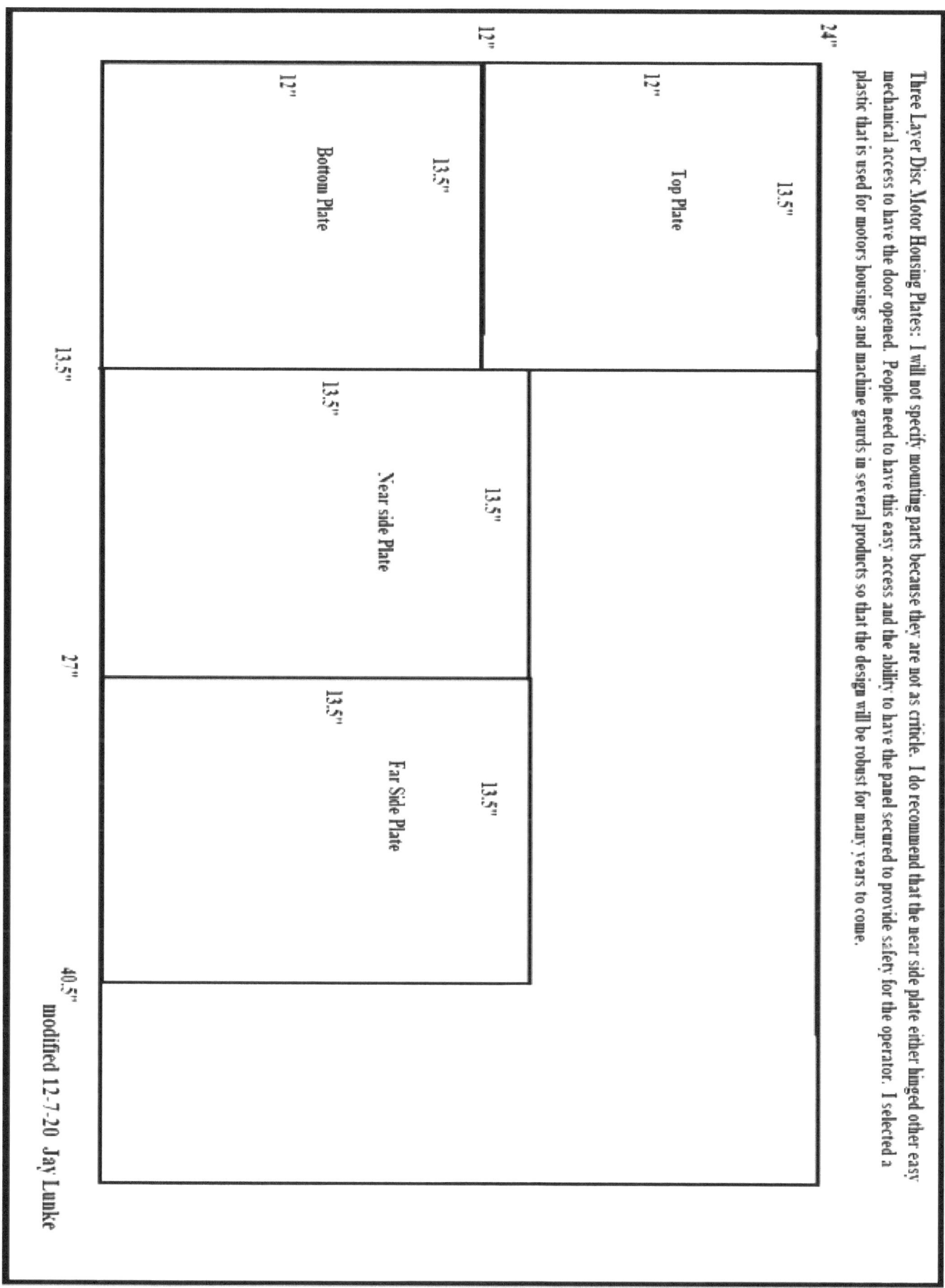

Three Layer Disc Motor Housing Plates: I will not specify mounting parts because they are not as criticle. I do recommend that the near side plate either hinged other easy mechanical access to have the door opened. People need to have this easy access and the ability to have the panel secured to provide safety for the operator. I selected a plastic that is used for motors housings and machine gaurds in several products so that the design will be robust for many years to come.

modified 12-7-20 Jay Lunke

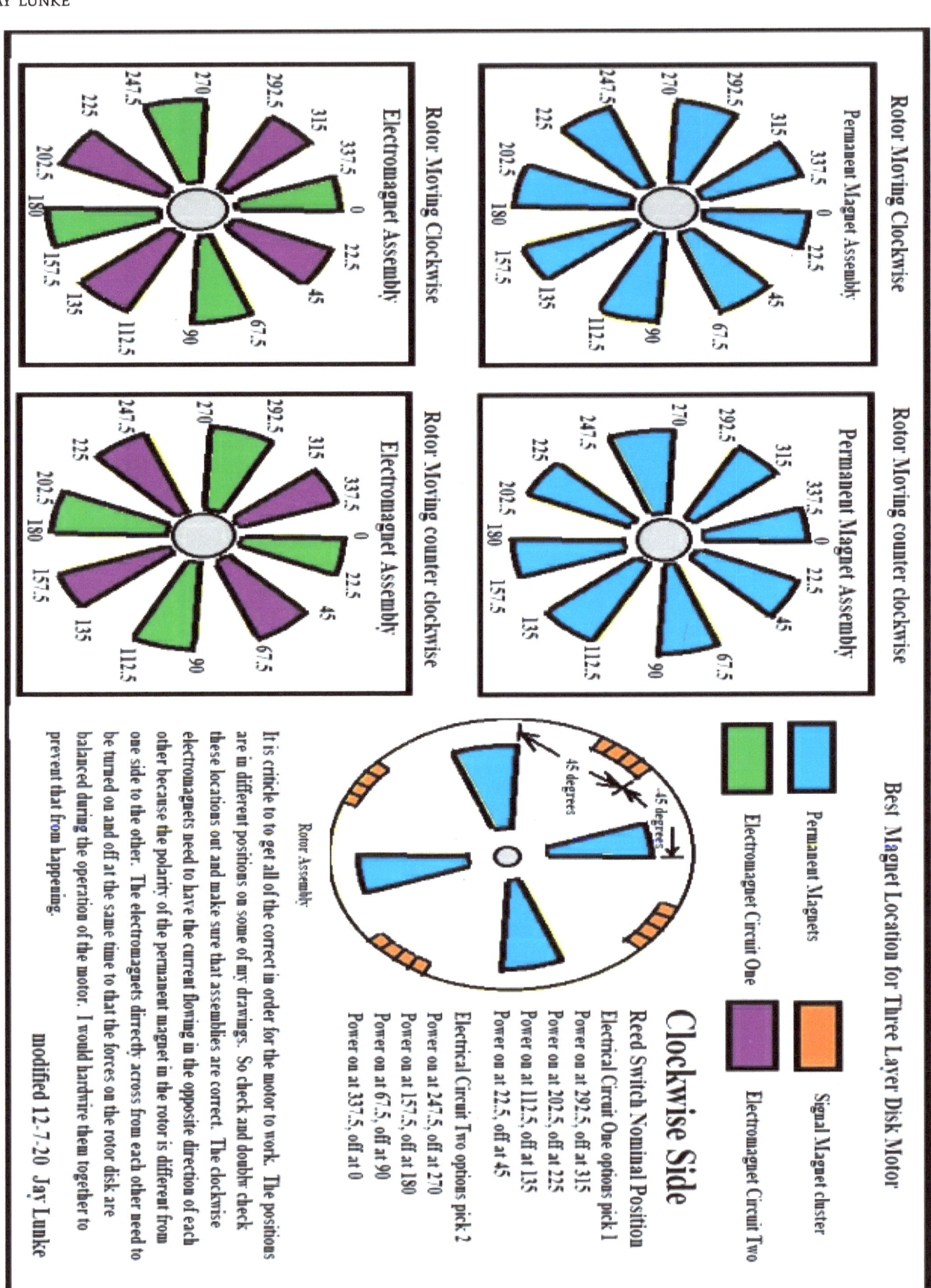

Three Layer Disk Motor parts list:

Note: The price may have changed since I found these parts on line.

1.) Wedge Magnet (8" outside Dia. 4" inside Dia. 22.5-degree circumference ¼" thick. Note: there are also optional ½" wedge magnets available for more money that can be used as well. There are several places that sell these wedge magnets. The lowest cost I found was about $10 a piece having blemishes on them. A total of 16 are needed.

2.) 2 each: Polyethylene sheets, 24" X 48" ¾" thick. These sheets can be cut up for the housing and plate assemblies. I found these for $102.76 each on Amaon.com

3.) Precision Shaft ¾" Dia. 24" long: I found them at Bearing Direct has them for $23.89

4.) Shaft Collar ¾" size bore. Zero has them for $3.38 each

5.) Thermoplastic Flange Bearing ¾" bore; VXB company has them for $17.79 each.

6.) Use Nylon hardware nuts and bolts and washers. Several companies have them.

7.) ½" threaded Nylon rod cut to 16" lengths. United State Plastic Corp. sells them in 4-foot lengths for $17.79. The proto-type needs 8 each at 16". So, buy 3 each 48" rods.

8.) ½"-13 Nylon nuts. United State Plastic Corp. sells a 100 each package of them for $14.70.

9.) 25-gauge motor wire to build the electromagnets. There are several places that sell copper motor wire. Other gauge sizes are also an option.

Now performing the ground work for this new technology can be very educational for the individual building and testing it. It is there where many more design options and improvements to the technology can be had. Consider building this prototype and go on from there. This new technology will not only help yourself, but it will help other people as well.

HOBBYIST CORNER:

When designing and building prototype devices, it is best to test out functions in the smaller sub-assembly levels first. This could save you a lot of time and money. I learned things the hard way. I will show you a design I built and failed later in this book. Yes, it was a learning experience and I learned things about magnets that I would not have learned before if I had not built the prototype. But before we get to that, the following drawing is a good tool that can be used in evaluating system components of a generator and motor system.

IN MY OPINION

If your goal is to build an over unity system, you should prove that it is self-running without any outside or hidden power source like an energized coil under the table or hidden battery in the wooden board the prototype is mounted on. Let me say it a different way. Having an open system is ok because it can be done. In my opinion, my motor designs operate in an open system where electrical energy is used to operate the electromagnets to produce torque. At the same time the open system allows the permanent magnets to contribute their torque to the design. The work that the motors perform can operate a highly efficient generator that in turn operate the electromagnets and power circuitry of the motor. The motor will still have torque to perform other work in the applications the motor was designed for.

The following drawing can test components of such a system without building a full-blown prototype. The device is a dual pendulum. It has two arm that are tied together. So, the motor component can be placed on one leg of the pendulum and the other leg of the pendulum can have the generator component on it. What you do then is to pull the pendulum legs up to 90 degrees and let it drop. As the drop goes though the other part of the design like a rotor moving through a stator assembly. As it goes through the structure you built at the bottom of the pendulum, an electrical pulse from the generator side is sent to the motor assembly arm. The pendulum, then swings up the other side. If the swing is greater than 90 degrees up, then you have achieved a COP>1. If not then you still have work to do. If I would have only listened to myself when I built the "All Permanent Magnet" prototype motor, I could have saved myself a lot of time and money.

This following drawing could save you a lot of time and money.

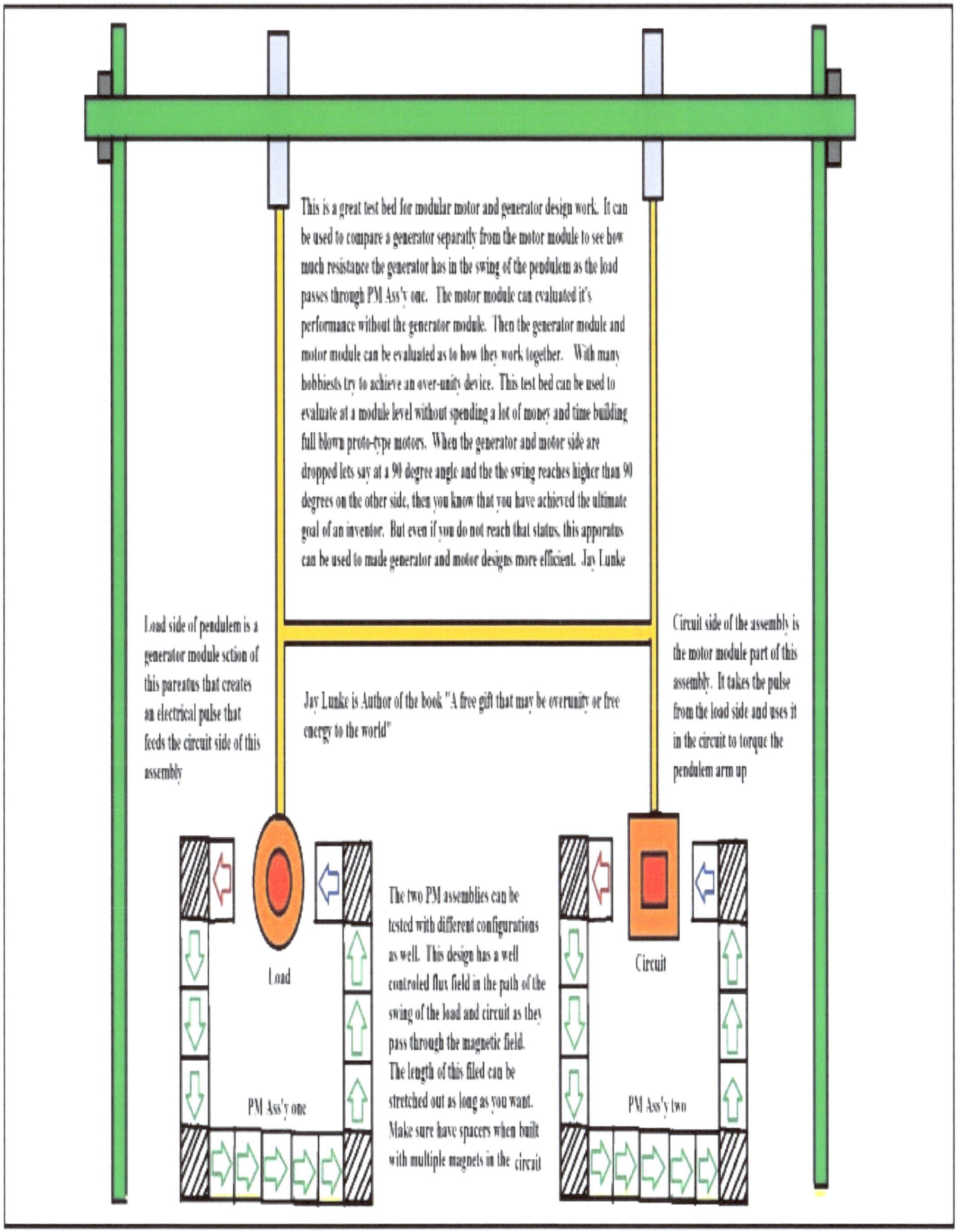

The base configuration can be anything you want. The drawing only has one example of what to use.

Now the follow drawing is a device I built for one of my old labs when I had them, in order to test shielding for motor designs. I did not find a material that could be moved freely between the magnets at the base and around the PVC pipe and cause the distance between the magnets to change. At least not in a way that I thought would be beneficial to my motor designs. I have not tested all of the materials yet.

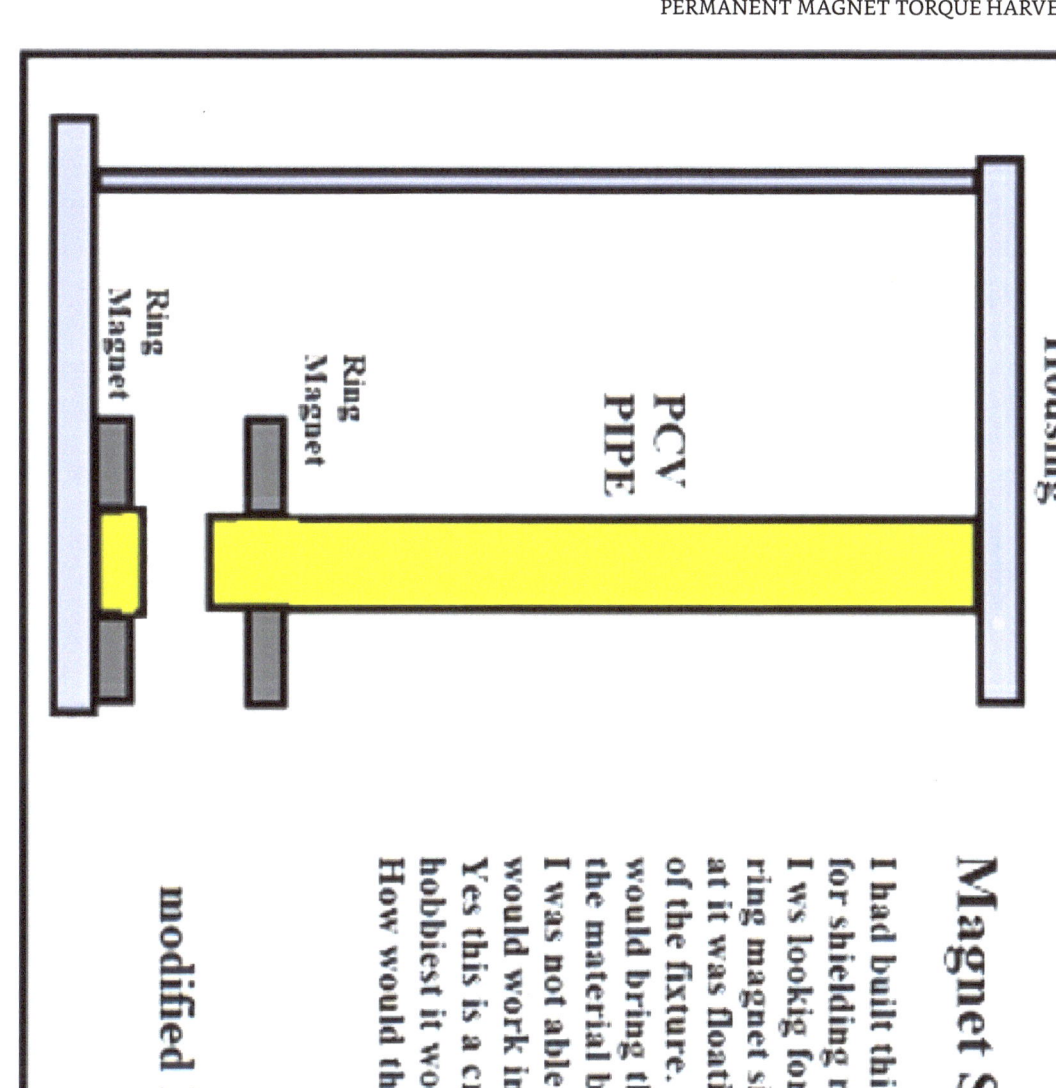

Magnet Shield Test Fixture

I had built this fixture to test different materials for shielding magnetic fields.
I ws lookig for a material that would cause the top ring magnet sitting on the upper PCV pipe to move at it was floating above the ring magnet at the base of the fixture. I was looking for a material that would bring the magnets closer together without the material being attracted to the ring magnets. I was not able to find a material that I felt that would work in a permanent magnet motor.
Yes this is a crude fixture, but for a blue collar hobbiest it worked fine for my needs.
How would the MU material work in this fixture?

modified 12-7-20 Jay Lunke

What I am telling you is that it is best to test out the unique functions that your device you want to build at as small of a level in order to save yourself a lot of time and money.

It is very hard to get someone else interested in building your designs when you do not have recognition in the field you are working in. It is best to find a way to do as much testing yourself. If you want to patent some of your work, in my opinion, you want to at lease have a working model first that a third source has performed an evaluation on it. The patent road is one that I have looked into, but have never done. But the percentage to patents that people have made more money than what they have spent is very small. So please be very careful with your resources. This again is only my opinion.

I am adding an All-Permanent Magnet design I had spent a lot of time and money on and how it failed me. I want you to feel the excitement I had before I built the motor through the words I had before building the prototype. Hopefully you can learn from my mistakes.

Failed All permanent magnet motor build,

Traveling Hybrid Flux Wave Motor

This motor name comes from the fact that the motor design is a re-configuring of the magnets on the stator assembly on the fly in order to create a moving flux wave in it continually in order to keep the rotor assembly move continuously. With the re-configuring, the rotor never crosses a transition point also known as a sticky point at any time during the motor operation. This motor design is unique for any all-permanent magnet motor ever built so far from the research I have done.

I have not built a proto-type motor using the "Three Layer Electromechanical Movement" technology yet but I am building one now after 50 years of working with designing motors. I have built partial motors using other technologies that I will not get into here. Without a working proto-type, I cannot claim over unity or free energy for my new technologies, that is why I say maybe it has a COP>1. I Have built test stations where I could test modules of the motor but what is needed is a fully operational motor for people to see. I am very excited to be able to be building a full-size proof-of-concept proto-type motor. I have chosen to build a motor design that would have no question of a COP>1 when I am done building it. As a picture is worth a thousand words, a working proto-type is worth thousands of words.

Before I start to explain the motor, I want to explain what this motor is similar to. Picture a man riding a train. The train is on a slope so that the gravity allows the train to move down the tracks. Now there is a limited number of tracks in front of the train. As the train moves down the tracks, the man picks up a section of track the train had just moved on and moves it to the front of the train so that it can move farther down the tracks. If there was no bottom of the hill this could go on forever because you never reach the bottom of the hill. What my motor design does is that the stator is constantly being reconfigured as the rotor is moved across it. The magnetism is like the slope on gravity; magnetism will move other magnets to certain positions to make an easier route of the flux to travel in the magnets. My design reconfigures the magnets on the stator assembly on the fly to keep the rotor magnets trying to line up with the stator magnets moving toward the sweet spot where the south pole of the rotor wants to meet the north pole of the stator magnet, but it never gets there. What my new technology does, is to create a forward torque between the rotor and stator through re-configuring the stator assembly through the full rotation of the motor travel. With the gravity driven train, it will reach the bottom of the hill, but with my train, since the magnetic poles keep moving because of the constant reconfiguring of the stator assembly, the rotor will never get to the pole it is trying to reach. The rotation of the rotor never ends because of the constant torque between the stator a rotor assembly caused by the continuous reconfiguration of the stator assembly. This book will go into a lot of detail of how this works in later chapters. I will only summarize what is happening with the technology here. This chapter gets more into the details of the proto-type motor and hardware I designed and am building. My greatest performing motors I have designed will not be used in this proto-type motor because those designs use electromagnets in the stator assembly in order to operate the re-configuration in them. Those designs do not need moving parts for the switching functions that needs to go on in the stator assembly for the reconfiguration of it. But with electric motors, laboratory testing is required in order to evaluate the efficiencies of them. This would be too costly for me to do. With an all-permanent magnet design without using electricity would be a visual test of its operation. There would be no doubt of its COP>1.

Now every permanent magnet and electro-magnet motor in the world have one or more flux waves in them. This is because it is the changing flux fields that create movement in them. With that being said most of those designs will have stationary PM and or EM components on both the stator and the rotor assemblies. The electro-magnets will have different currents and directions of currents flow through them in order to generate the changing flux to generate the torque to move the rotor inside of the stator assembly. These methods I will call two-layer systems. What I have been working with for the last few years are three-layer motor designs. What is different about three-layer designs is that the configuration of the stator changes while the rotor is rotating inside of the stator. This can be done either mechanically or electrically. If I do it electrically, usually I will use an air core for the electro-magnets or coils I am using in the motor. The reason I have been using an air core is because I want the component to disappear from the functional sight of the permanent magnets in the motor at certain times of the operation of the motor. It is because of the electromagnets appearing and disappearing functionally in the motor, I am able to change the functional make-up of the stator without moving hardware during the operation of the motor assembly. When the electro-magnet comes to life, it becomes a part of a hybrid magnet with the two adjacent permanent magnets next to it. This new hybrid magnet now has a different function with the permanent magnets on the rotor assembly. It is the cycling between the magnet to hybrid magnet arrangements that creates a condition where the rotor permanent magnets will always have a forward torque from the stator assembly to the rotor assembly. In this book, I use this method of design in many of my motor's designs. In my latest designs I have used physical moving parts in order to reduce the number of electro-magnets in the motor. I feel that the fewer electro-magnets I use then the less power will be needed to operate the motor. Now mechanical movement creates more losses that need to be taken into account as to the final performance of the motor. Once the technology is proven with mechanical movement, then the electrical motor designs will be explored with more enthusiasm and expectation of positive results in those motor designs.

Now what I have achieved by using the three-layer motor designs is to reconfigure the stator of the motor on the fly. When working with an all-permanent magnet design and trying to get continual movement between them there is that hump of "for every attraction there is a repulsion" between the two magnets that needs to be resolved. This is why in a lot of conventional electric motors there is a permanent magnet working with an electromagnet to control those negative torque points in the motor performance.

With the three-layer technology, let's say we are using two permanent magnets again, one in the stator and one in the rotor. Now let's assume that attraction is being used to create forward torque in my design. Now I would use the attraction of the two magnets to rotate the rotor half ways through the rotational travel from only the force existing between these two magnets, the magnets would now resist the movement of the rotor because of the repulsion between those two magnets. A conventional motor would have replaced one of the permanent magnets and used an electro-magnet in its place in order to provide a current in the opposite direction in order to move the rotor the rest of the rotor's rotation. What you have to do is have electrical energy in the electro-magnet through the full rotation of the rotor now. So, this means you have torque from 50% permanent magnets and 50% electromagnets. What the three-layer technology does is to reconfigure the stator into a functional magnet that creates forward torque from the stator assembly hybrid functional magnet and the rotor permanent magnet. This is the place where the normal repulsion occurs in the two-layer design. So, 50% of the rotation is done in the two-layer format and the other 50% of rotation is done with the three-layer format. So, you end up having torque from the permanent magnets and one hybrid functional magnet for the full rotation of the motor. The functional operation is through 50% of the rotation. When using the electromagnets in the design, you are operating at less power than other motors.

When using electrical energy, enough flux needs to be generated to create the hybrid magnet.

Now as the rotor moves inside the stator assembly using this alternating of hybrid to conventional configurations, then the device will have flux waves that are created and will move the rotor around in a circle during the operation of the motor design. Now there will be one flux wave for every four segments of rotor travel. If the motor has 32 segments to create one full rotation of the rotor, then there will be eight traveling waves in the motor.

Now for designs that use mechanical movement to make up the hybrid configurations, mechanical energy is needed in order for this operation to occur. The stator configurations need to be in sync with the rotor's rotation at the same time. Now there are endless ways to mechanically configure the hybrid and disassembly of it. What I have decided to use in my motor design is a switching wheel with eight segments of configuration on it.

PERMANENT MAGNET TORQUE HARVESTING

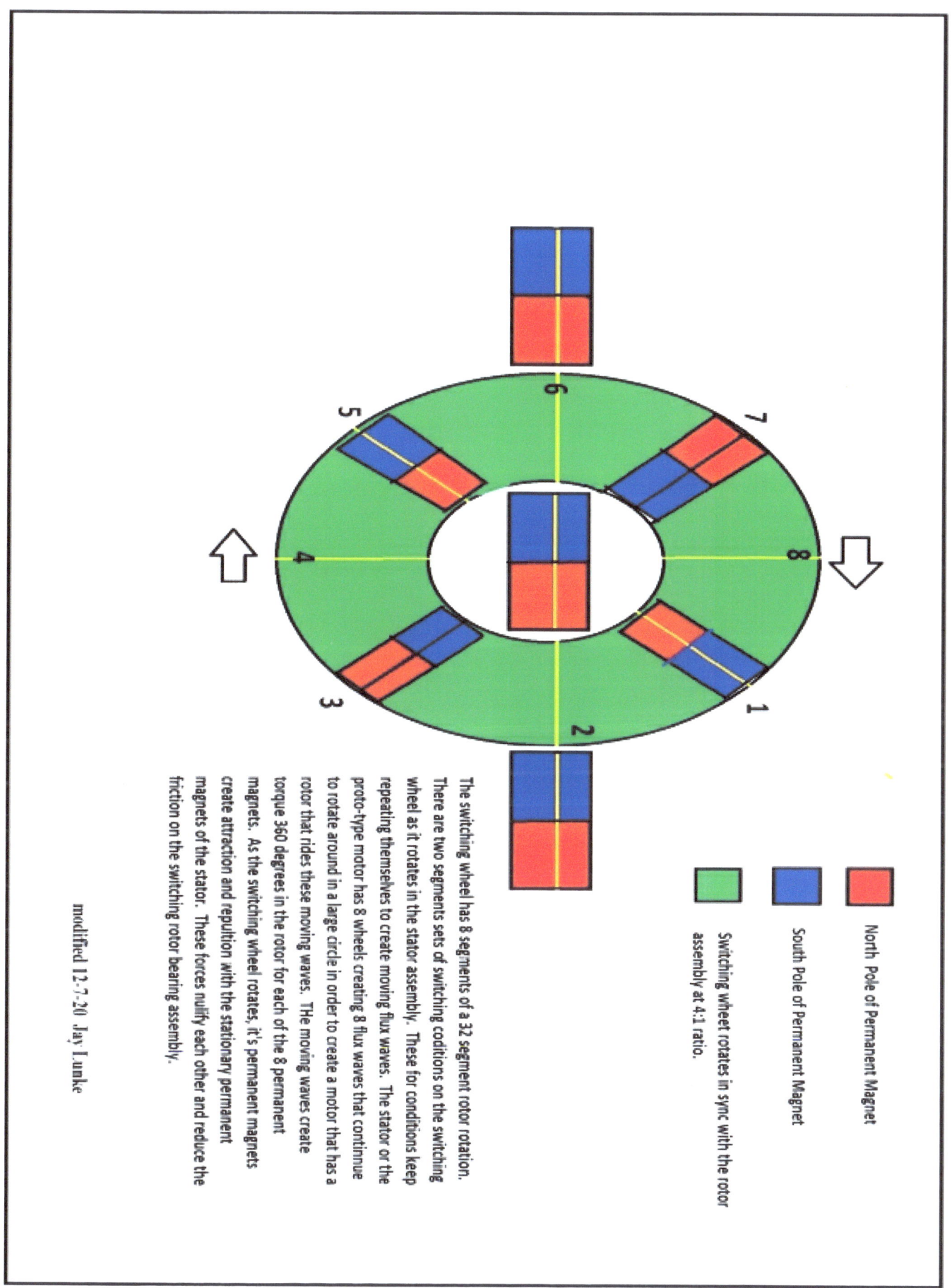

The switching wheel has 8 segments of a 32 segment rotor rotation. There are two segments sets of switching coditions on the switching wheel as it rotates in the stator assembly. These for conditions keep repeating themselves to create moving flux waves. The stator or the proto-type motor has 8 wheels creating 8 flux waves that continuue to rotate around in a large circle in order to create a motor that has a rotor that rides these moving waves. The moving waves create torque 360 degrees in the rotor for each of the 8 permanent magnets. As the switching wheel rotates, it's permanent magnets create attraction and repulition with the stationary permanent magnets of the stator. These forces nulify each other and reduce the friction on the switching rotor bearing assembly.

North Pole of Permanent Magnet

South Pole of Permanent Magnet

Switching wheel rotates in sync with the rotor assembly at 4:1 ratio.

modified 12-7-20 Jay Lunke

Each switching wheel works with a pair of switching positions along the stator assembly. I call one segment of travel as the length of one magnet on the stator assembly. The switching rotor is called out in segments of travel as well. There are four segment configurations on the switching wheel that are repeated again to create the eight segments on the switching wheel. Now what I have decided to use for my switching wheel is magnets for the switches. The switches will not only create the hybrid magnet on the stator assembly when needed to produce rotor travel but it will also place a magnet of the opposite polarity on the other end of the hybrid in order to strengthen the interaction of the hybrid with the rotor magnet. The way it is strengthened is that it helps to prevent the stator magnets into becoming one large ring magnet because of the stator permanent magnets close proximity with each other. By pairing two switching magnets going into switch positions having opposite magnetic poles to each other is a big advantage in reducing unwanted torque and friction on the switching wheel. As one switching magnet is being placed into the switching position it will have attraction to move into place. At the same time the other switching magnet will resist being put into place. The net result for the most part will be a cancelation on the switching wheel. Then when the first magnet is being removed from its switching position, it will resist that movement. While this is happening the other switching wheel magnet is being pushed away from the switching position. The forces between the two switching magnets again for the most part cancel each other out. Now the reason I said mostly cancel each other out is because if you perform a vector analysis of these movements, there will be some overall resistance to performing this function along with friction in the bearings, wind resistance of the spinning switching wheel and other misc. performance reductions that become a factor. Now in the final motor assembly, the rotor magnet will also contribute some additional reduction to the performance of the switching wheel.

A series of the switching wheels can be added together to create a track for linear movement. What I do is to bend my switching assembly for the rotor to move in an arc. I put 8 of these assemblies together so that the eighth section connects to the first switching network again. I end up with a high torque rotational motor design. This works good for the traveling flux wave to have a continuous path to travel.

PERMANENT MAGNET TORQUE HARVESTING

The Stator operates like a wave pool pushing 8 flux waves around in one big circle as the 8 switching wheels rotate to generate these waves.

The rotor permanent magnets have forward toque on them through all of the rotation of the rotor as a result of this action like a person surfing on an ocean wave or a dog chasing its tail.

The electromagnets can either operate at a 50% duty cycle for forward torque when needed for the motor or by switching the dirrection of the current through the coil at the correct time to generate forward torque throughout the rotation of the rotor assembly.

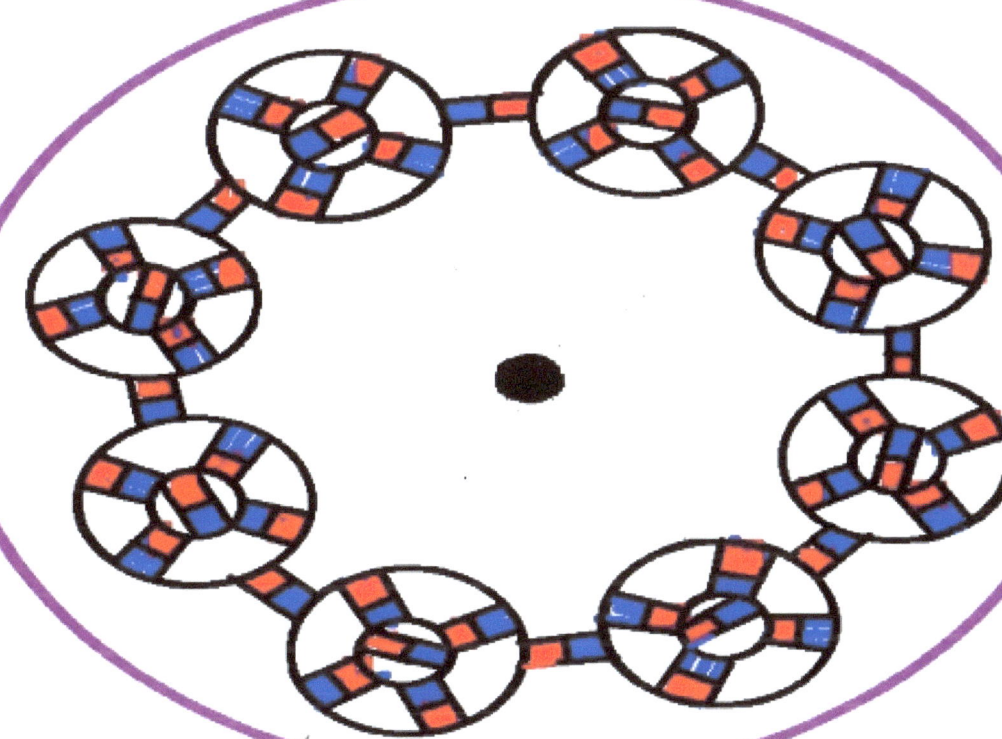

Stator Assembly

The trade-off; The more the stationary magnets are lined through the switching magnets, the better the ballance in the switching wheel will be. The farther away the switching magnets are from the stationary magnets in the off condition, the better the performance of the motor.

WHen the functional magnet becomes larger, the less torque is between the rotor and stator magnets will be.

Eight flux waves are created and will flow around the large circle in sync with the rotor movement because of the gears are driving the switching wheels

modified 12-7-20 Jay Lunke

Now as the rotor moves along the tract of the stator there is always torque between the stator and the rotor assemblies supporting the rotation of the rotor. Now the rotor is connected through a gearing system to each of the switching wheels. Since the stator has 32 segments in one full rotation of the rotor and the switching wheel has eight segments of function on it, the switching wheels will rotate four times around for each rotation of the rotor. So, the four to one gearing ratio between the rotor and switching wheels will create a one-to-one segment movement between the stator and rotor assemblies. The segment positions between the rotor and switching wheels are critical. So that is one reason there can be no slippage of the main shaft to the rotor disk, main shaft gear to main shaft, switching wheel to switching shaft and the switch wheel gear to switching wheel shaft. Since there will be torque on these components, then extra design thought needs to go into them. Now the main 6" gear has 192 teeth in it. The next 3" gear has 96 teeth in it. The 1 ½" gears have 48 teeth in them. This means that gong from the 6" gear to the 1 ½" gear will have a ratio of four to one. The more teeth the better the synchronization will be because there will be less gear slop in the motor. I used 17 gears instead of 9 gears in order to save on cost for this proto-type build. It just so happens that the extra layer of gears causes the rotation of the rotor and switching assemblies to be in the same direction. Functionally, this should not make any difference.

Since a picture tells a thousand words, I am adding pictures of the proto-type here to show how the rotor and stator can work together for this new technology.

PERMANENT MAGNET TORQUE HARVESTING

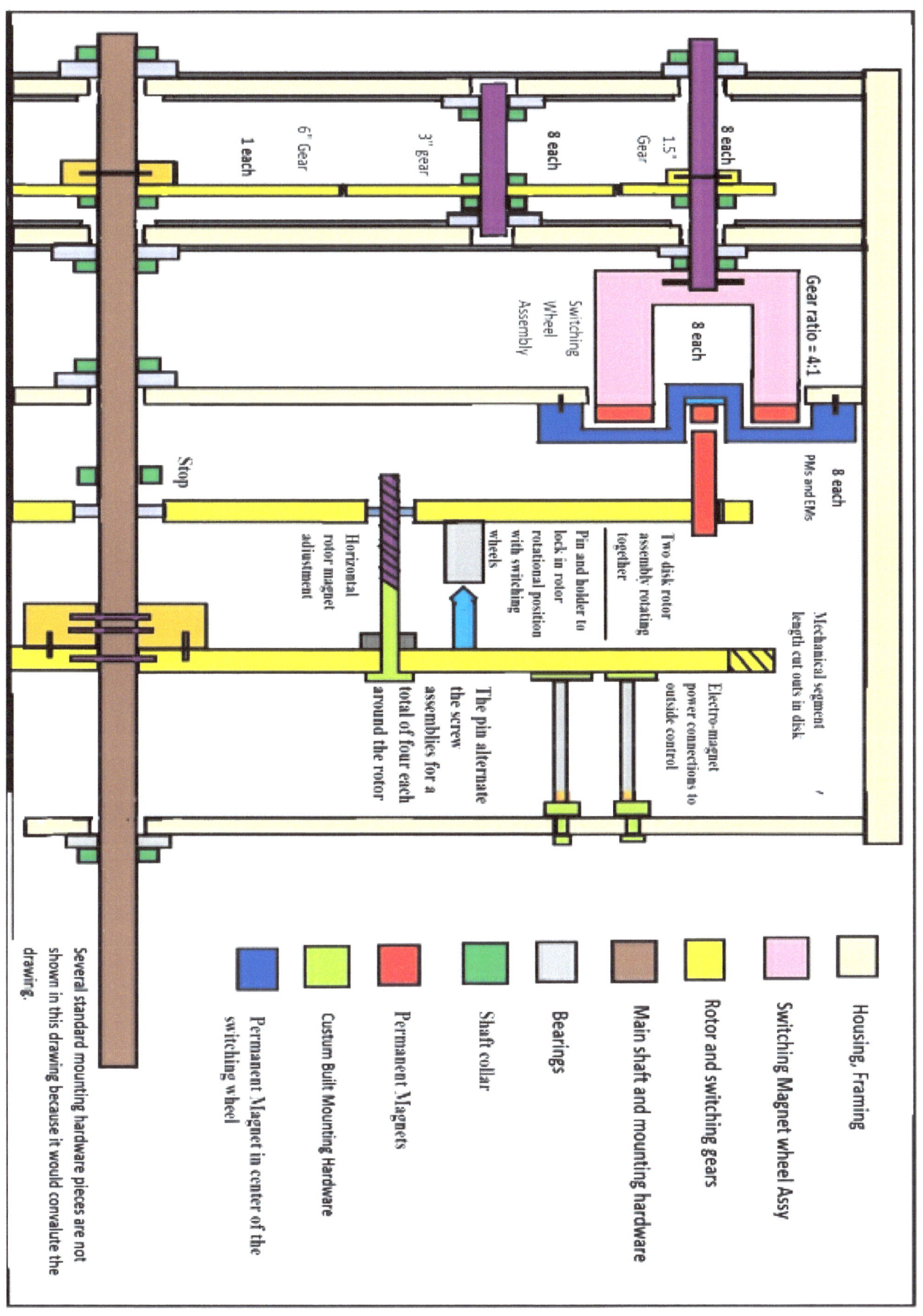

107

This drawing shows connection for an external power supply to drive the optional electro-magnets if needed to power the motor

Electromagnet Value to motor

EM1 has the most benifit to the motor

EM2 and EM3 have less additional benifit and will not be built into the functional proto-type motor.

When using dual power supplies, the current can be switched through the electro-magnet in order to produce forward torque in the motor through the full rotation of the rotor assembly.

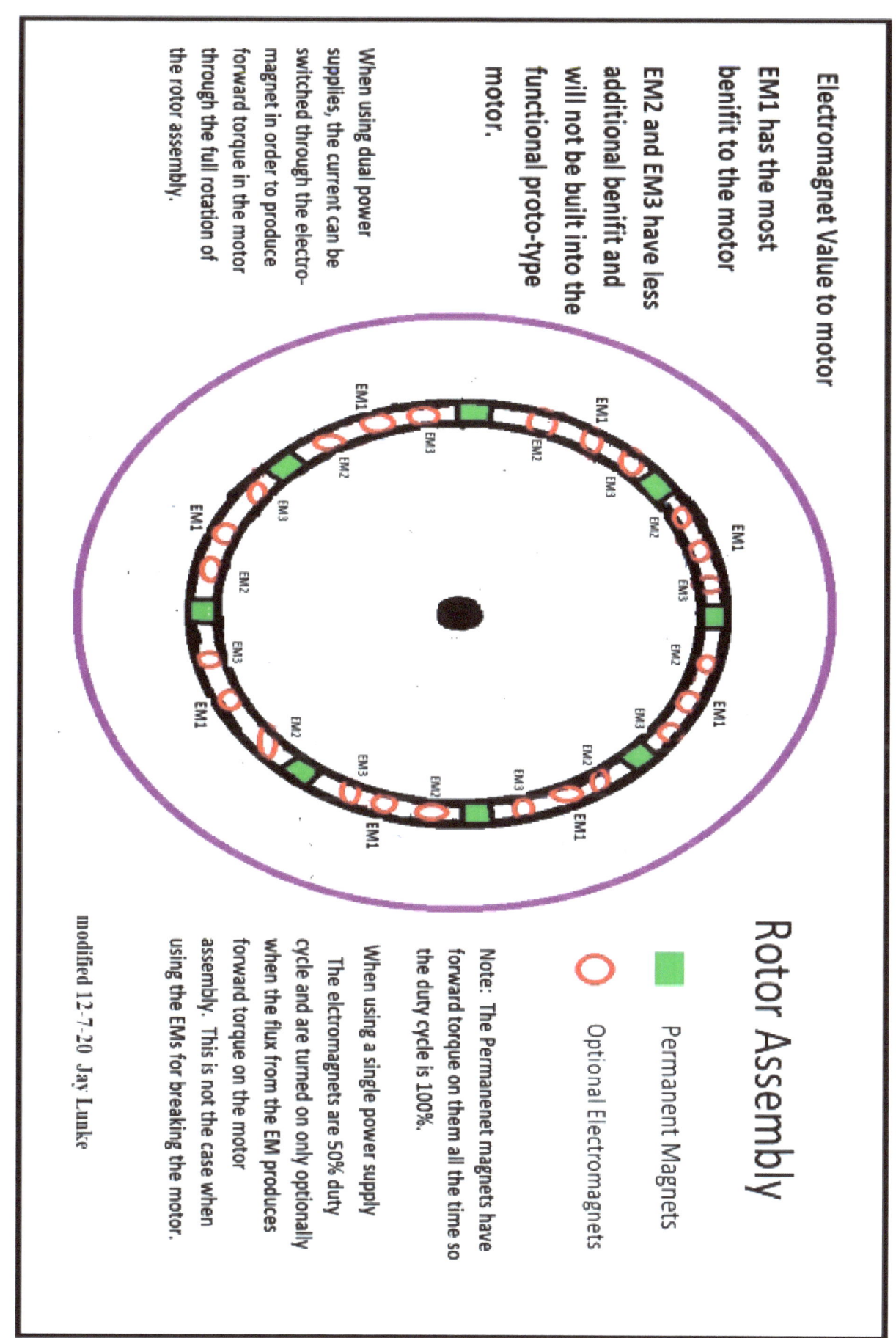

Rotor Assembly

■ Permanent Magnets

○ Optional Electromagnets

Note: The Permanenet magnets have forward torque on them all the time so the duty cycle is 100%.

When using a single power supply The elctromagnets are 50% duty cycle and are turned on only optionally when the flux from the EM produces forward torque on the motor assembly. This is not the case when using the EMs for breaking the motor.

modified 12-7-20 Jay Lunke

Since this proto-type will be able to test a few different concepts with it. This drawing shows the rotor makeup of permanent magnets and electromagnets.

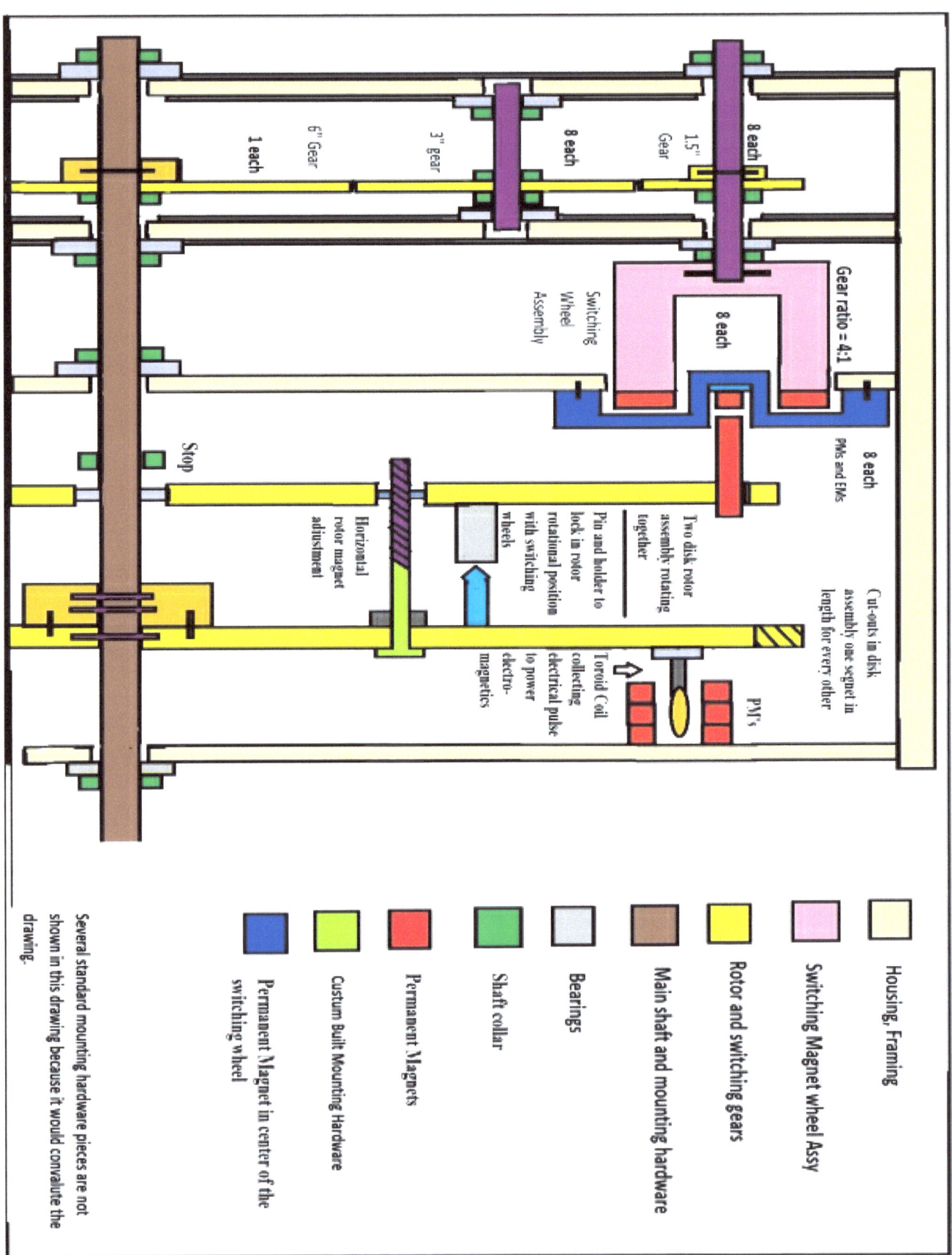

This drawing above shows the motor and not an on-board generator.

Now the design of the on-board generator could be easily installed in the top section of this assembly because of all of the extra room for one.

Now the dual pendulum could be used to optimize the best on-board generator for someone to power the electromagnets that are optional to install on this motor design to turn the rotor assembly. The final proto-type will have both an on-board generator and connections for external power supplies to operate those electromagnets. Yes, the first proto-type will be kind of a test platform. Even the rotor is built with two layers to adjust the height of the rotor magnets. All of these things will help to optimize the design for the next level of proto-type that would require a machine shop to build it.

Now even if the all-permanent magnet version operates the motor, I want to add the electromagnets to increase the power of the motor. With the way and amount of power is injected into the electromagnet, the motor can have a wide range from a stopped position to a fast rotation. The rotation cannot be two fasts because the switching wheels would fall apart at higher speeds. This limits the applications for this motor design.

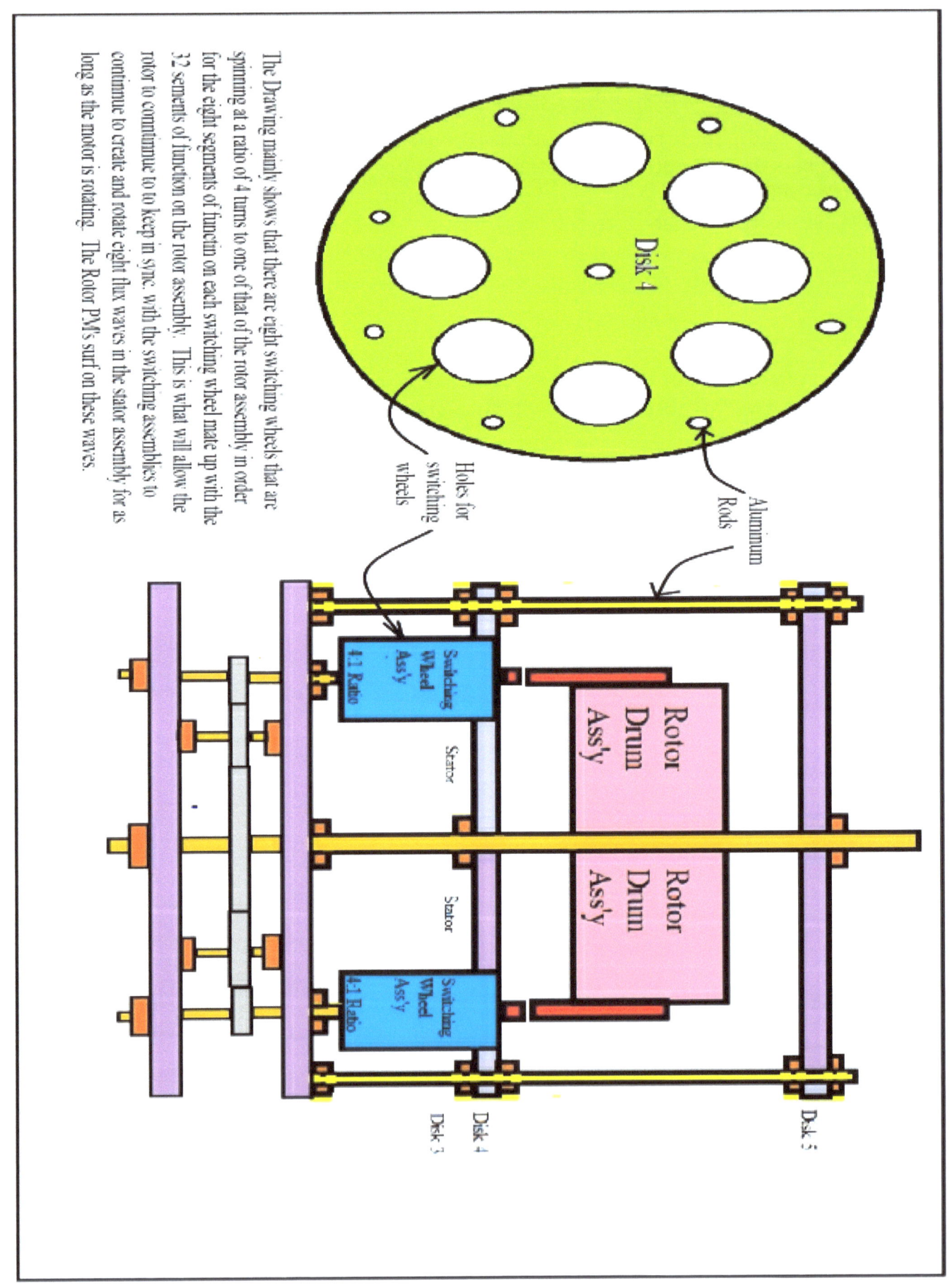

The Drawing mainly shows that there are eight switching wheels that are spinning at a ratio of 4 turns to one of that of the rotor assembly in order for the eight segments of functin on each switching wheel mate up with the 32 sements of function on the rotor assembly. This is what will allow the rotor to continnue to to keep in sync with the switching assemblies to continnue to create and rotate eight flux waves in the stator assembly for as long as the motor is rotating. The Rotor PM's surf on these waves.

This drawing shows more information of how the hardware and framing are built in the proto-type. (not to scale)

I have added four photo's I took with my laptop. These photo's will reveile a lot of poor workmanship on my part. A lot of cuts in the aluminum parts were done with a hand saw. Plumbing PCV pipes were used in the switching wheels. I had to rework the switching magnet wheels three times because the power of the magnets were way stronger than the flexing of the switching wheels could withstand. I had to file the openings for the switching wheels larger because of it. I am not a machinist. I only wanted to prove the operation of this configuration but instead I disproved it. So if you try to build an All Permanent Magnet motor, please do not try this one, you will be greatly disapointed.

But it is amazing what I learned from this build. Before I built this prototype, I was still at the 3:1 permanent to electromagnet ratio. Now I am at a 5:1 PM to EM ratio. I added the disk motor prototype design and modified my tank circuit to incorporate the reed switches and switching logic. I believe that it was also during this time frame that I added the steering diodes in the modified tank circuit in order to allow a lot more slop in the capacitor value that is used in the tank circuit. I added the notation for using a starter switch in order to bring the motor up to speed before letting the tank circuit to take control. I also have had new options for the power generation circuits. I brought my "flow through motor" technology into the Three Layer Electromechanical Movement Technology through a redesign of it. I found out that the switching wheel does switch in and out permanent magnets with the attraction and repulsion forces to equalize them. This design function is a major contributer in some of my motor designs in the book I wrote named "The core to fee energy". The switching wheel along with the introduction of the magnetic field from the rotating rotor the way I designed my prototype motor caused most of my designs failure mode. The new design I use in my new book should not have that problem.

So maybe the failed prototype turned out to be a great experience for me after all.

So much can be learned from having a hands on build and test. For those who have not done it yet, I strongly recommend that you do it.

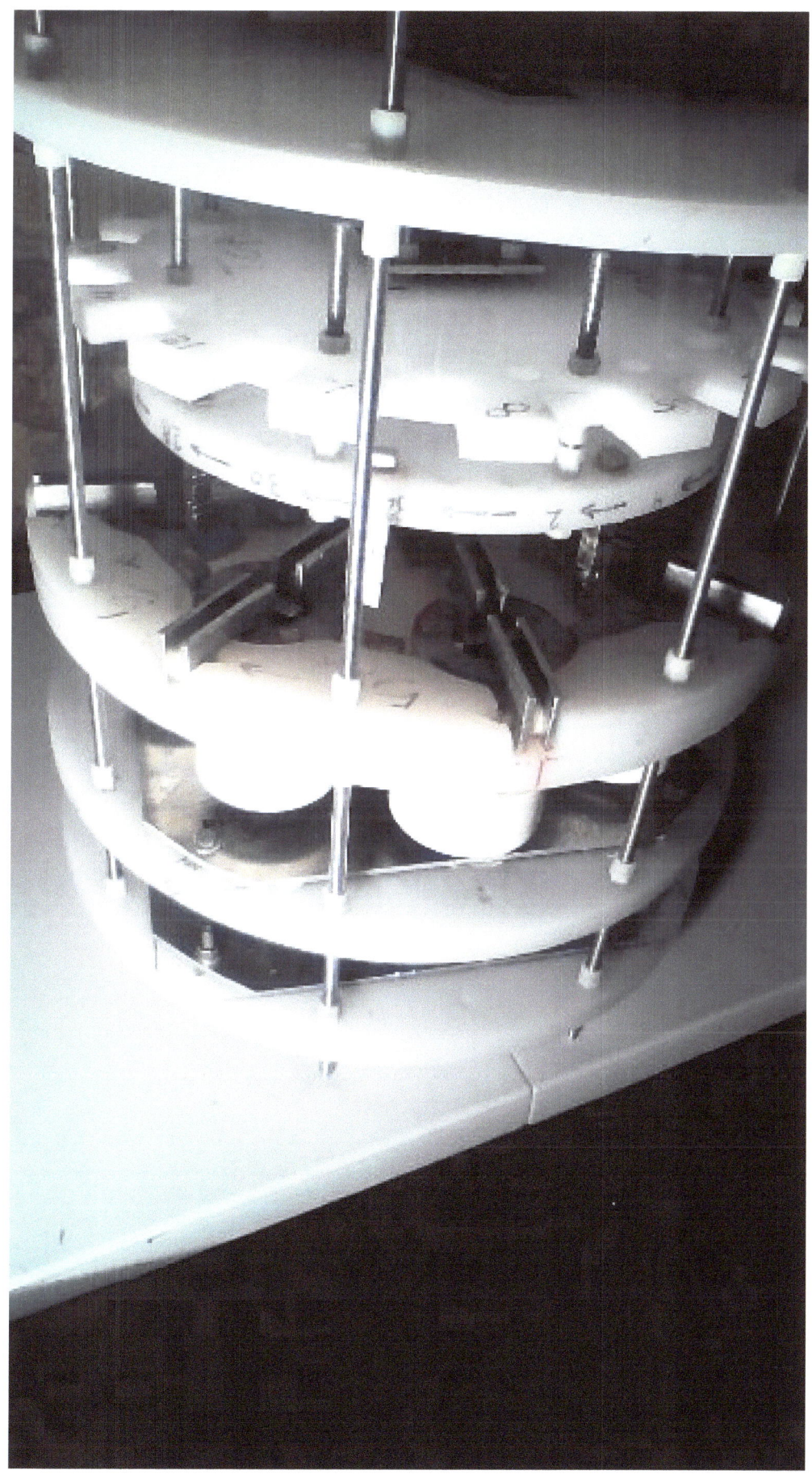

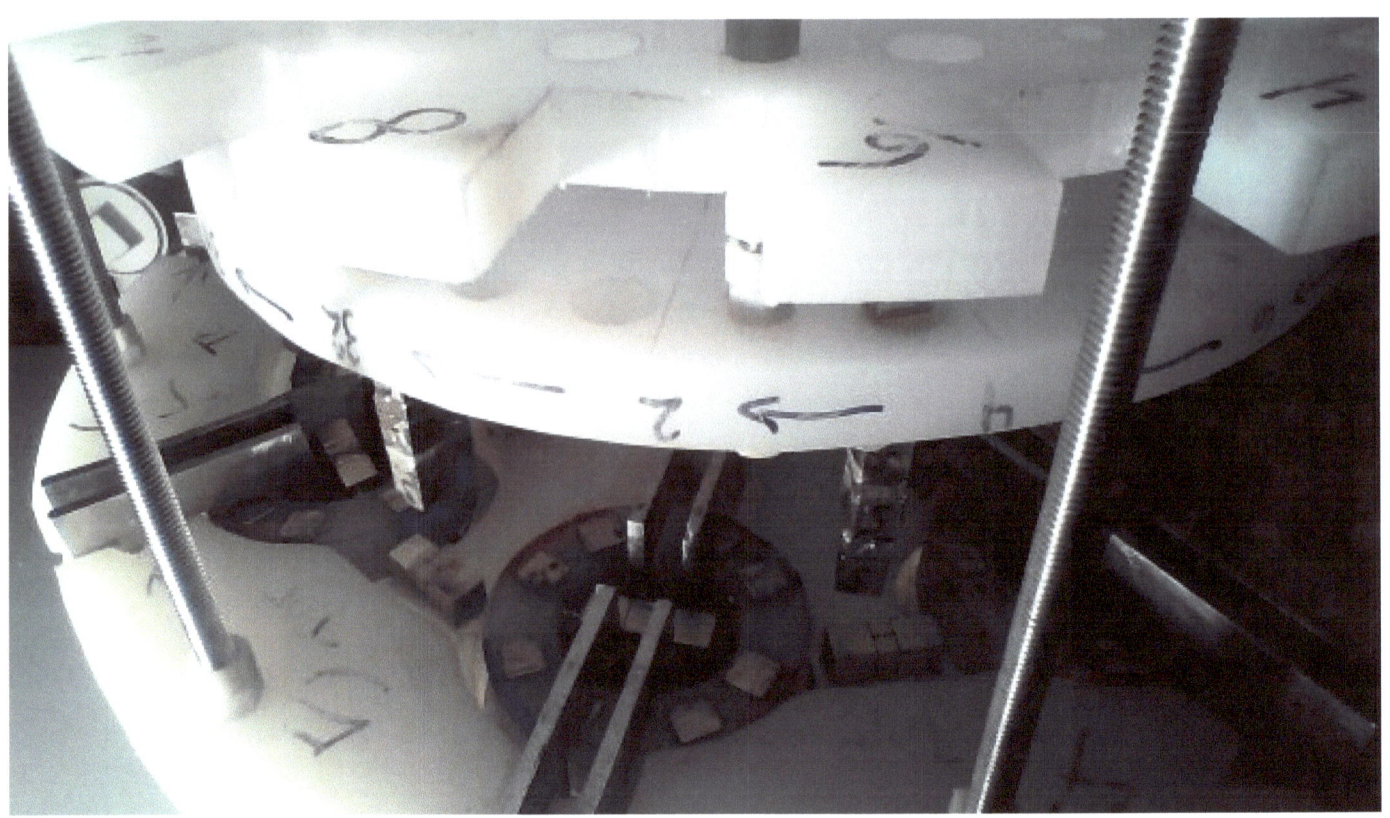

JAY LUNKE

Whenever I do designs, I am always wondering is there anything else I can do to improve the design. Sketch those ideas out. Those sketches can bring out more ideas and so on. So, I will show some thoughts I had before my prototype failed.

This drawing below only has the basics to show that a common gear box can be used to drive two motor sections. The polarity on the stator would need to be switched on one of the motors in order to have the torque on the rotor in the same direction. (note to scale)

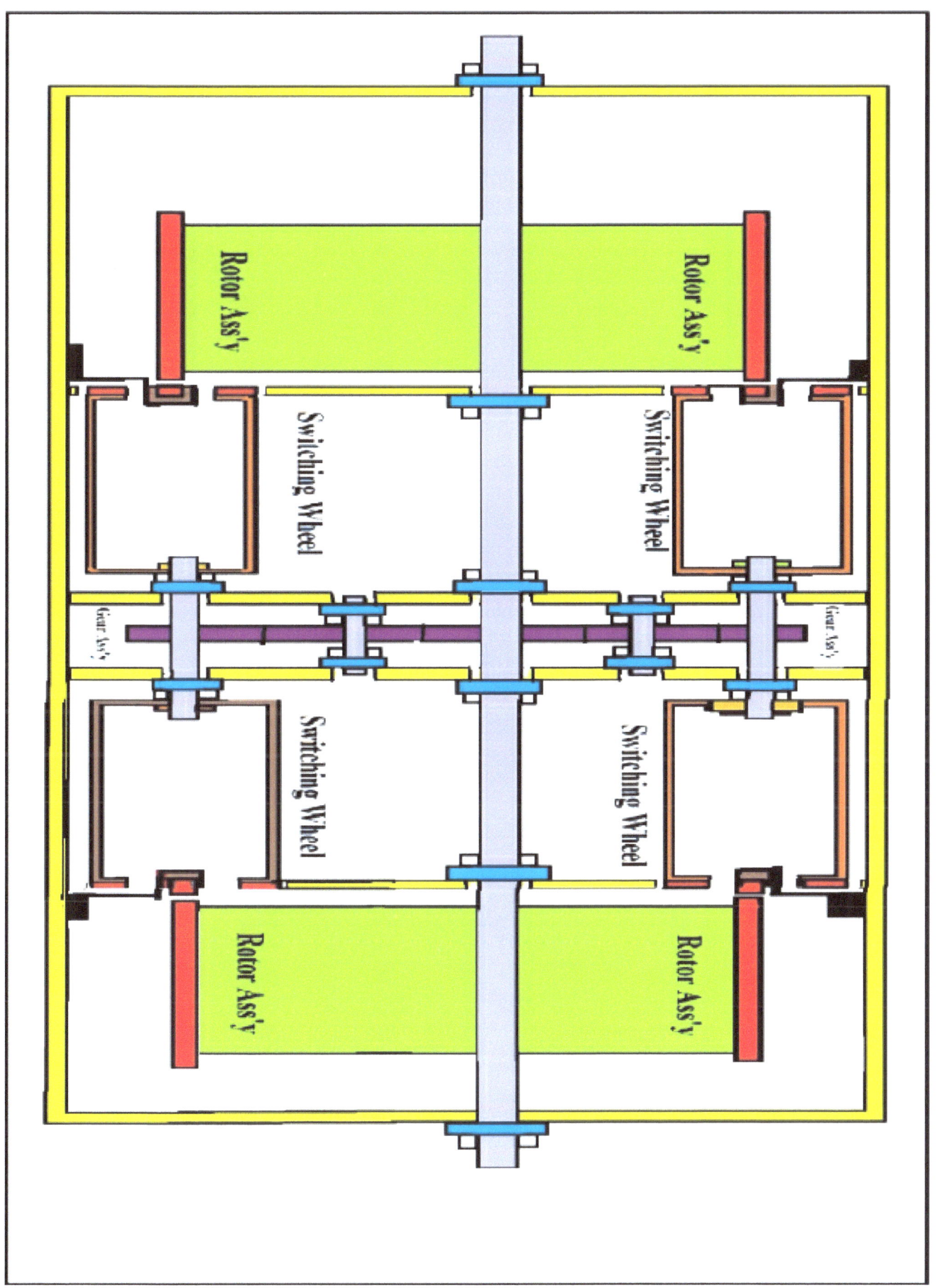

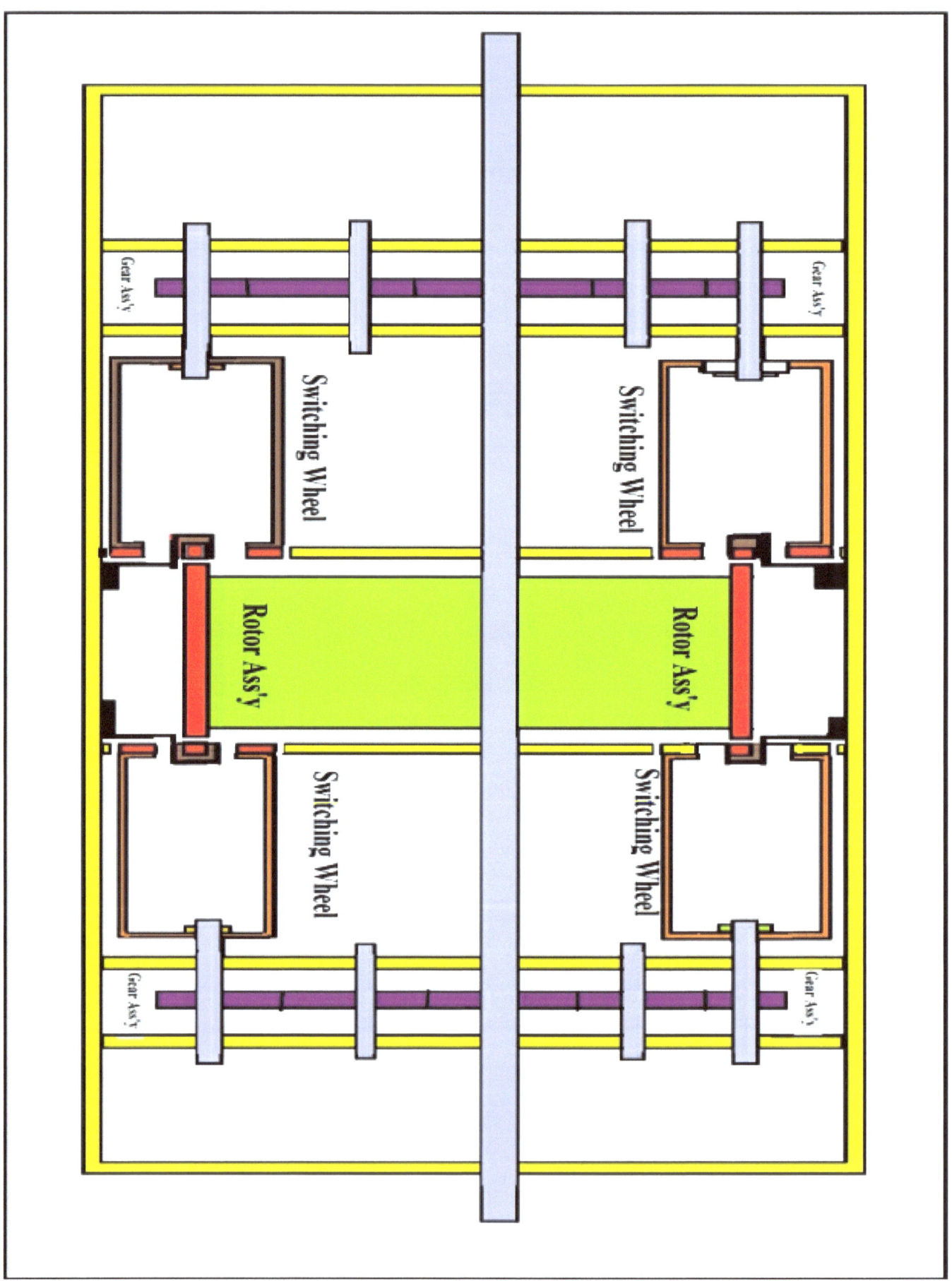

This is another option of using a common rotor assembly. This saves on permanent magnets needed if two motors were built to increase power output of the motor. (not to scale)

Now with all of the options shown above, they all have a traveling wave in the stator assembly that the rotor travels along the sweet point of that flux wave generating the movement of the motor. In the following drawings do not let the backwards arrows deceive you as to resistance to the motors movement because it builds the strength of the forward torque of the flux wave. These drawings are meant to provide you with a better understanding of how the flux wave is built and moves in the stator assembly. I get a lot more in drawings, pictures and charts than what I get from a lot of words. That is why I have so many of them in this book.

The rotor PM's ride on the angular green bars of this graph. Between the green bars the electromagnets take advantage of cycling the power of the electromagnet to keep forward torque on the rotor throughout the full travel of the rotor assembly.

The graph following the next graph is more technical because it summarizes the torques and resistance to the movements of the switching assemblies as they apply to the final overall performance of the motor assembly.

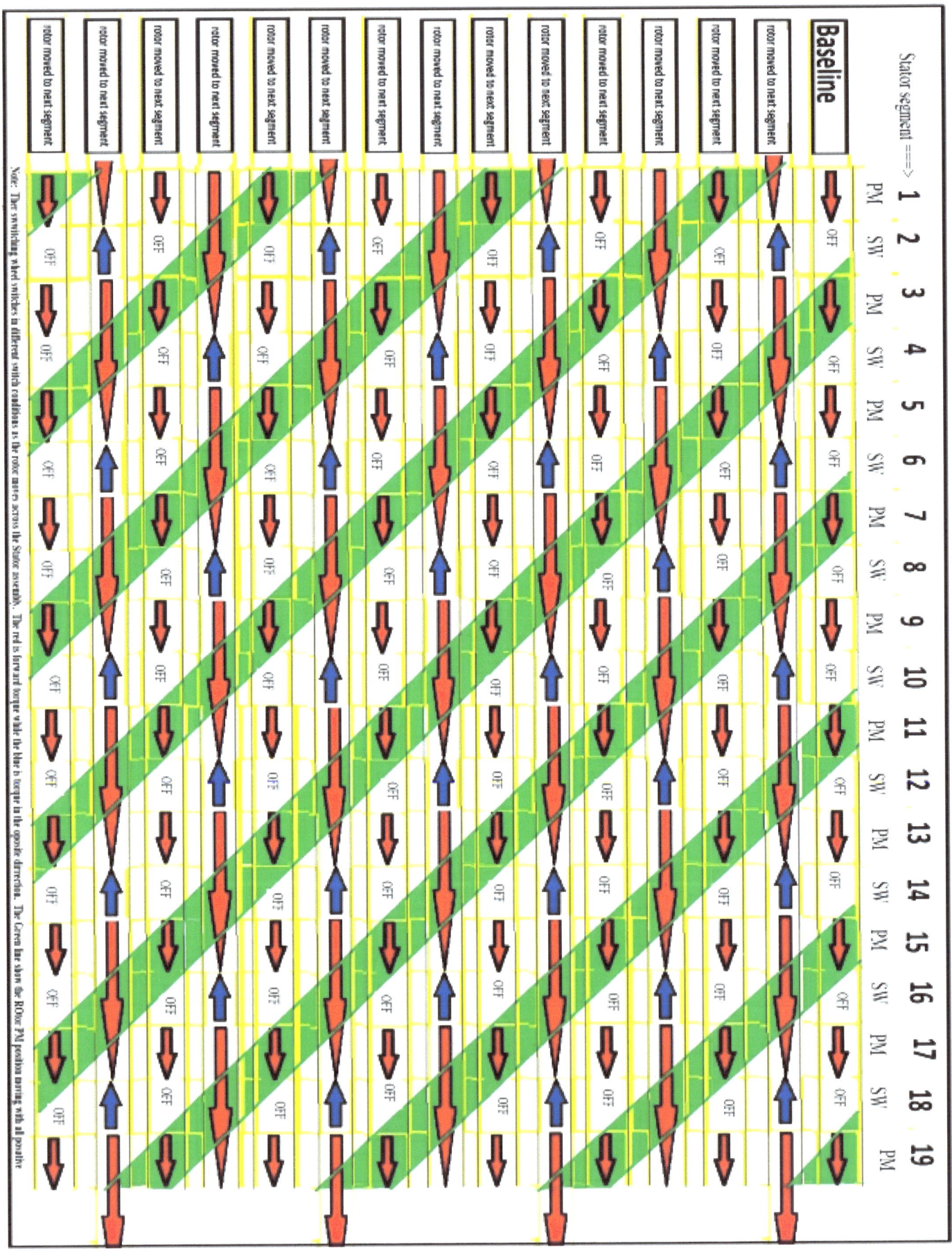

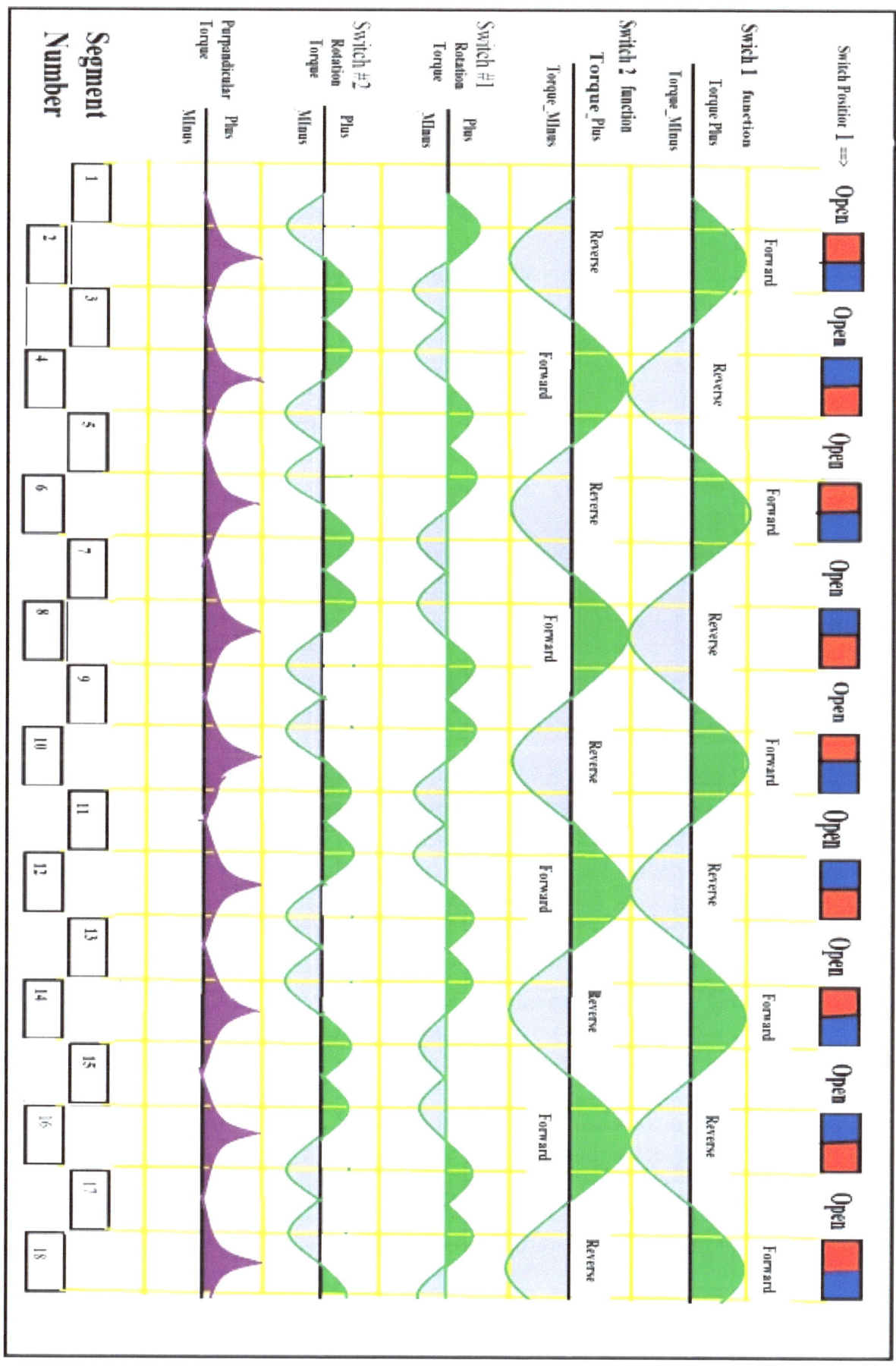

The top two lines on the drawing show the function of the switches in the stator assembly that are used in generating the traveling flux wave in the stator assembly. The following three lines on the drawing show the forces upon the switching wheel assemblies themselves. These switching wheels to my knowledge have not been used in motor designs before.

Now for the big question? How does the continuous torque the rotor has to turn the rotor compare to the switching wheel losses? If the losses are less than the rotor forward torque, then the rotor will have continuous movement. This is before I add the electro-magnets to the rotor assembly. Since creating a functioning motor on permanent magnets alone is low, building the motor with the alternating electro-magnets into the motor is a must for the proto-type design. Now with that being said I will be looking to the electro-magnets being fed from an on-board generator. The on-board approach will allow me to reduce the hardware and circuitry than the conventional way of having a separate motor and generate to operation in motor in a test system. IF and that is IF the electrical pulses the generator generates on the on-board generator is enough power to power the electro-magnets in the rotor to create continuous movement, then I would have a COP>1. This is why I am designing this into the proto-type before building the motor. Even if the on-board generator works on this motor design, the motor would have to have a way to bring the motor up to speed. One option is to start the motor with another motor and once operational, then the external motor could be removed from the circuit. Now another option would be to drive the electro-magnets from an external power supply until the motor comes up to speed and then it could be removed.

Now this prototype had a lot of build challenges by themselves. The main shaft is too thin that causes a lot of flexing between the rotor and switching wheels. With electromagnets added to the design, it is now not an all-permanent magnet design for this type of motor.

There are some problems I ran into with my prototype build.

I did not have dowel pins that were snug in the assemblies. Because of this, I added a lot of glue. Anywhere there was a magnet mounted on a smooth surface, it would break off during rotor rotation. I not only had to rough up the area, but also drill little holes for the glue to seep into for extra holding strength. The switching wheel would violently swing from side to side as the switching magnets passed by stator magnets. This would cause vibration that would tare the switching wheels apart for sure at lower speeds. There is so many more things that were a struggle.

Now there may be a way to perform the moving the switching magnet into and out of the area that the rotor passes around the stator assembly perpendicular to it during the operation of the motor assembly. I would start out with a 5:1 PM to EM design because then you will have more torque to begin with that the losses of the switching magnets will need to overcome for the motor to be non-functional in my opinion. Also, I would change the magnet orientation to that of the disc motor because the magnets have the most interaction at the poles of the magnets. I would also think about increasing the segments of travel to at least 64 segments of travel so that the functional magnets do not become too big. But I personally will be working on generator concepts next. But with the claims I have seen out there of no torque generators, this may be useless work to be involved in, I guess we will see.

The following drawings are some optional ways for the switching to occur. There are for sure many other ways to do it as well. The switching magnet being a permanent magnet, will not work as long as the resistance to the movement of this magnet into and out of switching position are more than the torque produced in the rotor to rotate in the motor assembly. I can not forget that the rotor movement will contribute to the switching magnet movement as well. I know other people have been and are working on reducing magnetic movement resistance for other devices. I was even wondering if the way the magnetic bearing operates could somehow be incorporated into the design. Well, these are just a few options to look at.

All Permanent Magnet Motor

This is a modification of the all permanent magnet motor I built that is redesigned to reduce the resistance in the switching wheels rotation. It also changes the orientation of the stator permanent magnets to increase the torque between the rotor and stator assemblies. The old switching wheels had four magnets, two in the positive direction and two in the negative direction. The new wheel has one magnet followed by an open space. This configuration will create eight traveling flux waves in the stator assembly.

This two changes will reduce the switching resistance at the same time as produce forward torque between the rotor and stator assemblies. The big question is will the torque generated in the rotor in this motor design be greater than the resistance to the turning switching wheels. This is something I do not know how to model. So if you have modeling software, could you model this motor design?

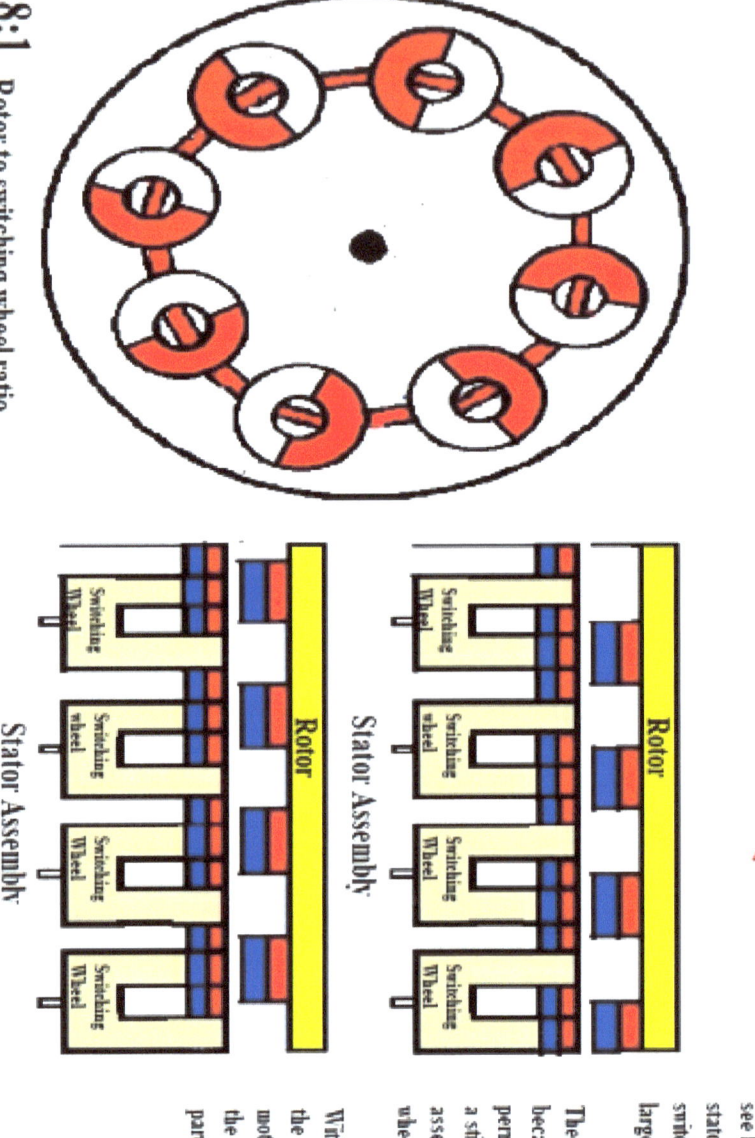

8:1   Rotor to switching wheel ratio

You need to go to the drawings already posted about this motor design to see how the rotor permanent magnets can rotate perpendicular to the stator assembly that has the stationary permanent magnets along with the switching wheels. From the lessons learned, I would change to much larger shafts, and reduce the number of gears from 17 down to 9 gears.

The big reason I believe this design needs to be built and tested is because this motor uses the Three Layer technology that keeps permanent magnets between the rotor and stator assembly from having a sticky point. It does this by constantly reconfiguring the stator assembly during the rotation of the rotor in the motor. The switching wheels perform the reconfiguration need to achieve that objective.

With the switching wheel now has the orientation of the magnets in the same direction, the resistance will be a lot less than the older motor design. The attraction of going into the switching position with the repulsion of leaving the switching position should for the most part cancel each other out.

modified 12-7-20 Jay Lunke

PERMANENT MAGNET TORQUE HARVESTING

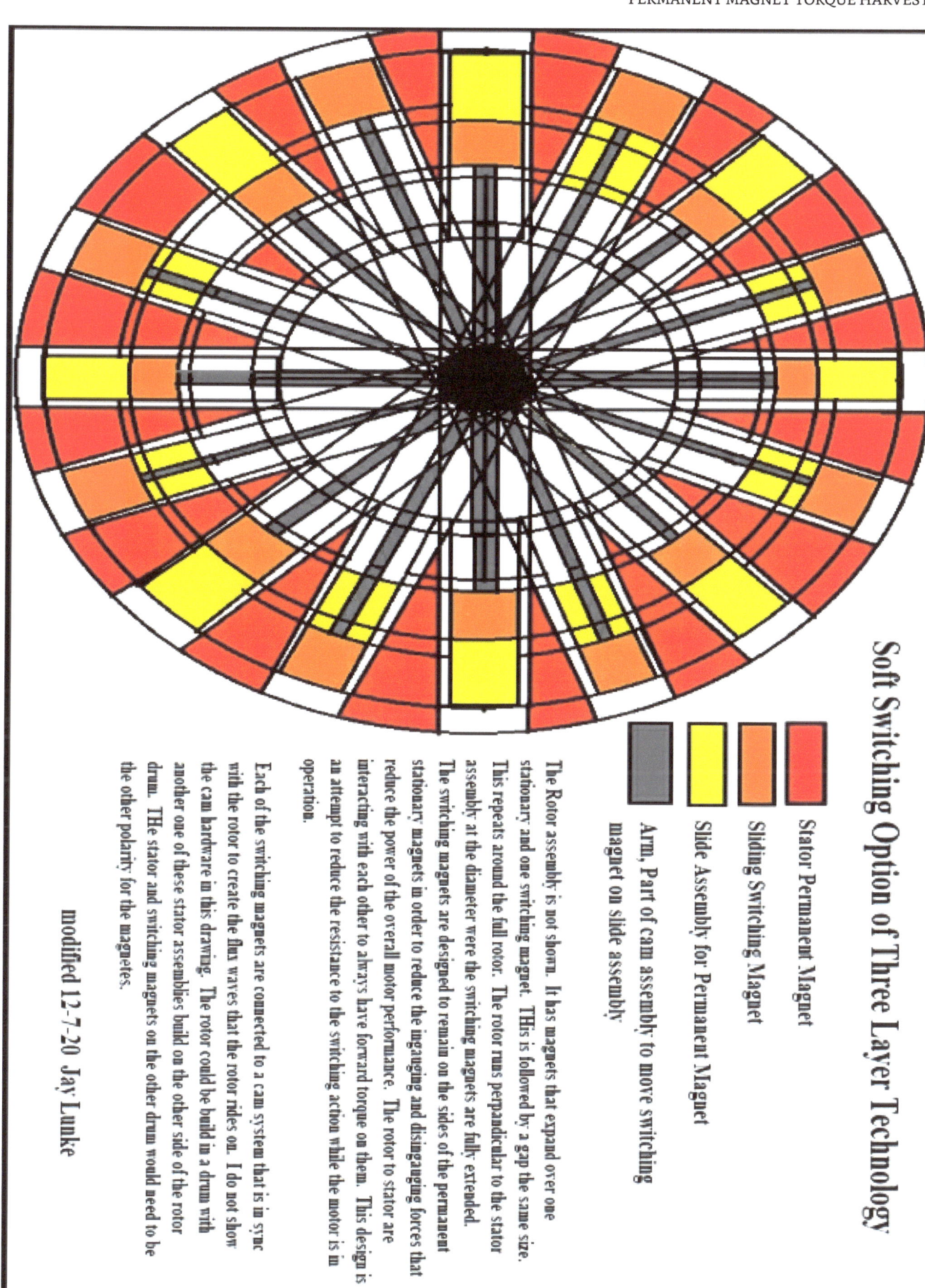

Soft Switching Option of Three Layer Technology

- ■ Stator Permanent Magnet
- ■ Sliding Switching Magnet
- ■ Slide Assembly for Permanent Magnet
- ■ Arm, Part of cam assembly to move switching magnet on slide assembly

The Rotor assembly is not shown. It has magnets that expand over one stationary and one switching magnet. THis is followed by a gap the same size. This repeats around the full rotor. The rotor runs perpendicular to the stator assembly at the diameter were the switching magnets are fully extended. The switching magnets are designed to remain on the sides of the permanent stationary magnets in order to reduce the ingauging and disingauging forces that reduce the power of the overall motor performance. The rotor to stator are interacting with each other to always have forward torque on them. This design is an attempt to reduce the resistance to the switching action while the motor is in operation.

Each of the switching magnets are connected to a cam system that is in sync with the rotor to create the flux waves that the rotor rides on. I do not show the cam hardware in this drawing. The rotor could be build in a drum with another one of these stator assemblies build on the other side of the rotor drum. THe stator and switching magnets on the other drum would need to be the other polarity for the magnetes.

modified 12-7-20 Jay Lunke

129

All Permanent Magnet Motor

This is a modification of the all permanent magnet motor I built that is redesigned to reduce the resistance in the switching wheels rotation. It also changes the orientation of the stator permanent magnets to increase the torque between the rotor and stator assemblies. This configuration will create eight traveling flux waves in the stator assembly.

The yellow lines show the path of the rotor peremanent magnets.

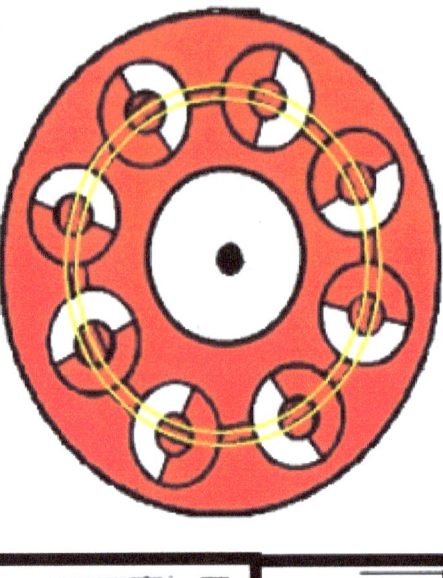

8:1 Rotor to stator switch ratio

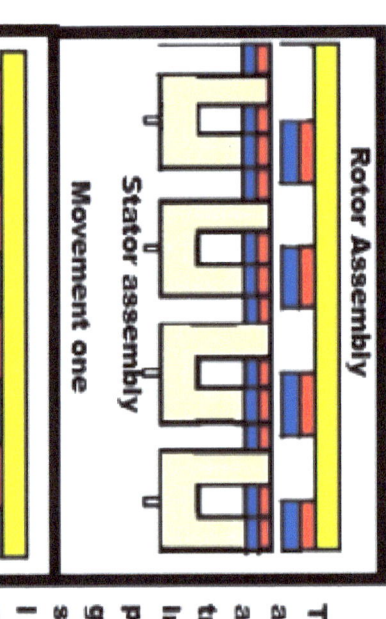

The stator cycles between movement one and movement two. In theory the rotor assembly should rotate along with the traveling flux wave.
In order for this to work, the torque produced by the traveling wave needs to be greater than the torque to rotate the switching wheels.
I have not built a prototype of this design. This design is simpler to build than the prototype I built and provide photo's of in my book.

Modified 1-11-2026 Jay Lunke

Now if you can come up with a way of performing the switching movement with less resistance than the torque created in the rotor assembly, you will have a great over unity motor on your hands.

By sharing my failed attempts to build an All-Permanent Magnet Motor, you may find that some part of the design that may make sense to you. You can make your own modifications to build your own prototype motor. Maybe your design will be successful.

ABOUT THE AUTHOR

I have been an inventor since I was a child. My twin brother is the same way. Both of us would do and build things that we had never seen other people do before. I am not going to share all those stories here. But the greatest interest I have had is what makes a magnet work. In high school I purchased several permanent magnets and would work with them in several ways trying to figure out how to use them in linear movements. Even though all of the engineering books say that permanent magnets can not by themselves produce continuous movement, something told me that there is something that must have been overlooked about magnets because I can feel the power that they have when I play with them.

Once I learned about electromagnets, I learned that they worked like a light switch that turns the lights on and off in a house. I thought that their must be a way that this characteristic should be able to be used in conjunction with permanent magnets in order to bring out usable work from the permanent magnets. At first, I worked with what I call the two-layer (rotor to stator) assemblies to try to create the most efficient motor. This is where I came up with the flow through motor designs 40 years ago. I include the work of these motors in my book "A Free Gift That May Be Over Unity or Free Energy to The World". This book has a lot of information about different motor configurations and applications using that technology.

When I tried to patent this technology. I ran into some issues that caused me great difficulty if I designed and patented the technology. It caused me to stop designing motors for several years. I did develop the "Three Layer Electromechanical Movement in the 1980s during that time, but kept it to myself. Then one day a man named Lee at work asked me if I was still working on motor designs. He said that there was a lot of people working on different concepts on the internet. So, I started looking and I started working with my design work again. It was because of my bad experiences in the past and some of my spiritual believes that I decided that instead of tying to get rich on the work I have done with motors, I want to freely share that work with other people. I want everyone to be able to benefit from my work. So, I further developed the new technologies and wrote another book named "Permanent Magnet Torque Harvesting". But as great as I think these books are, I wrote another book with my wife Paula, that is about the spiritual journey that is so much greater than these two books. In case you have not noticed from this book, I see things differently from everyone else. In the book "Please Pass the Pew", I share some things about the spiritual life that you will never find in any other spiritual book. The Bible is the worlds best book and I am trying to live by the teaching of it. What does it benefit Jay, if Jay should gain the whole world, but Jay loses his own soul? If I patented all of my work, the money I would receive from it would rune me. So, I sincerely ask you to please read my book "Please Pass the Pew" before other books, except the Bible. It will be a book that will challenge you more than any secular book you have read before. All of my books are sold on Amazon.

Jay Lunke

CONCLUSION

If it where possible, and I think that it is, people would want to build an All-Permanent Magnet Motor. But in reality, most of these motors would have lower power to them than the current magnetic motors on the market today. These motors would not be able to be used in most vehicles. With the harvesting torque from permanent magnets into a motor design using the Three Layer Electromechanical Technology design could produce motors that could be superior in efficiency to the current magnetic motors used in vehicles today. With this new technology along with the modified tank circuit with steering diodes could meet the over unity status.

This new technology does not create energy from nothing but is harvesting the torque from permanent magnets. The tank circuit adds more conversion stages of the energy that is already in the system to make sure the highest efficiency is used in producing torque from the electrical energy that is used to power the electromagnets.

Just think of the fossil fuels that can be conserved if this new technology can harvest energy from permanent magnets.

DISCLAIMER

The information in this book is theory. The designs have not been built and tested unless stated otherwise. I have not done a patent search to see if my ideas are unique. If you use the ideas in this book, you need to do your own patent research. I am not responsible for any legal expenses for people taking and using these ideas. I welcome anyone to freely use the information in this book after they have searched for patents and gone through the other legal avenues necessary to do so.

With 80,000,000 patents granted in the world, I cannot say that someone may have come up to the same conclusions as I have in their own research. Even if I did, someone could come up with new discoveries after I have written and published my idea's.

Since my theories have not been tested, all of my work is my opinion. My opinion can and has been wrong as I will show you an example of that in this book.

If I had unlimited time and resources, I would have built a lab and tested all of the devices I have in this book. I am an old senior citizen with limited resources. I am not responsible for the validity of the contents of this book.

www.ingramcontent.com/pod-product-compliance
Lightning Source LLC
Chambersburg PA
CBHW051149220526
45473CB00003B/714